Processing and Conversion of Oil and Gas: Modeling, Control, Simulation and Optimization

Processing and Conversion of Oil and Gas: Modeling, Control, Simulation and Optimization

Editor

Jean-Claude Assaf

MDPI • Basel • Beijing • Wuhan • Barcelona • Belgrade • Manchester • Tokyo • Cluj • Tianjin

Editor
Jean-Claude Assaf
Chemical and Petrochemical Department
Saint Joseph University & Lebanese University
Beirut
Lebanon

Editorial Office
MDPI
St. Alban-Anlage 66
4052 Basel, Switzerland

This is a reprint of articles from the Special Issue published online in the open access journal *Processes* (ISSN 2227-9717) (available at: www.mdpi.com/journal/processes/special_issues/oil_conversion).

For citation purposes, cite each article independently as indicated on the article page online and as indicated below:

LastName, A.A.; LastName, B.B.; LastName, C.C. Article Title. *Journal Name* **Year**, *Volume Number*, Page Range.

ISBN 978-3-0365-6723-5 (Hbk)
ISBN 978-3-0365-6722-8 (PDF)

Contents

About the Editor

Jean-Claude Assaf

Dr. Jean-Claude Assaf is an Associate Professor at the Lebanese University and a lecturer at the Faculty of Engineering and the Faculty of Sciences at Saint Joseph University. Dr. Assaf has won several national and international awards including LIRA (Lebanese Industrial Research Achievements) award. His expertise lies in chemical and petrochemical process design, simulation, and optimization. Dr. Assaf is also the owner of several industrial patents and has published numerous articles in peer-reviewed journals. Since 2018, Dr. Assaf has been recognized as outstanding reviewer for a number of high-impact journals in which he conducted successful reviews. Furthermore, he edited different books and was invited as guest editor in international peer-reviewed journals.

Preface to "Processing and Conversion of Oil and Gas: Modeling, Control, Simulation and Optimization"

Processing and Conversion of Oil and Gas: Modeling, Control, Simulation, and Optimization describes the core of the activity of petrochemical engineers, process controller engineers, industrial engineers, energy engineers, and researchers in the oil and gas field.

Petroleum is considered the black gold of the earth, but this treasure cannot be utilized without the usage of innovative and up-to-date technologies for its recovery and conversion. To the best of our knowledge, the crude oil extracted from wells is not in a pure form and is mainly composed of oil, gas, and water. Thus, the need for an oil processing plant to separate it from other fluids and impurities in an environmentally friendly manner is crucial. Therefore, these remaining contaminants can be highly problematic during oil transportation or usage, hence novel optimized methods are needed to improve oil processing.

Natural gas is nowadays considered the main bond between existing fossil fuels and future resources. Nevertheless, it may contain low amounts of chemicals, such as hydrogen sulfide, carbon dioxide, nitrogen, and helium. Therefore, natural gas processing is designed to refine the raw gas by eliminating impurities, thus producing a dry natural gas compliant to the pipeline gas quality specifications. Natural gas processing plants can remove several types of contaminants, such as water, hydrogen sulfide, carbon dioxide, mercury, and other hydrocarbons. The application and improvement of different operations including sweetening and dehydration make the raw natural gas ready for usage or conversion.

The conversion of oil and gas is mainly conducted through refineries and chemical processing plants. Hence, oil and gas refining leads through the combination of several physicals, chemical, and thermodynamical processes to the formation of many valuable products, including fuels, materials, chemicals, and/or heat and power. To note that even the polluting emissions of the refineries that can be formed at any stage of the oil or gas conversion process have a certain economic value and can be further processed or sold. Moreover, the impurities removed during the oil and gas processing step are not always considered as waste but instead, they are considered as potential by-products of this industry. Therefore, by aiming toward a green process and limiting energy consumption, economic and environmental benefits will take place.

This book explores the modeling, control, simulation, and optimization of new and revamped petrochemical processes using proven software. Thus, it offers novel illustrative examples, prospective applications, and solutions to improve these processes. Furthermore, applying numerical methods and optimization at both the theoretical and practical levels are within the scope of this book.

Jean-Claude Assaf
Editor

processes

Editorial

Special Issue on "Processing and Conversion of Oil and Gas: Modeling, Control, Simulation and Optimization"

Jean Claude Assaf [1,2,*,†] and Marina Al Daccache [1,†]

1 Ecole Supérieure d'Ingénieurs de Beyrouth (ESIB), Saint-Joseph University, CST Mkalles Mar Roukos, Beirut P.O. Box 11-514, Lebanon
2 Faculty of Engineering (ULFG), Roumieh and Hadath Campus, Lebanese University, Beirut P.O. Box 6573-14, Lebanon
* Correspondence: jeanclaude.assaf@net.usj.edu.lb
† These authors contributed equally to this work.

Citation: Assaf, J.C.; Al Daccache, M. Special Issue on "Processing and Conversion of Oil and Gas: Modeling, Control, Simulation and Optimization". *Processes* **2023**, *11*, 388. https://doi.org/10.3390/pr11020388

Received: 11 January 2023
Accepted: 17 January 2023
Published: 27 January 2023

1. Introduction

Petroleum is considered the black gold of the earth, but this treasure cannot be utilized without the usage of innovative and advanced technologies for its recovery and conversion. To the best of our knowledge, the crude oil extracted from wells is not in a pure form and is mainly composed of oil, gas, and water. Therefore, an oil processing plant to separate it from other fluids and impurities in an environmentally friendly manner must be employed. The residual contaminants can be highly problematic during oil transportation or handling, hence novel optimized methods are essential to improve oil processing. Nevertheless, natural gas is nowadays considered the main bond between existing fossil fuels and future resources. Natural gas is primarily composed of methane and may contain variable amounts of other higher alkanes. Besides, natural gas may have low amounts of chemicals such as hydrogen sulfide, carbon dioxide, nitrogen, and helium. Natural gas processing is designed to refine the raw gas by eliminating impurities, thus producing dry natural gas compliant to the pipeline gas quality specifications. Natural gas processing plants can remove several types of contaminants such as water, hydrogen sulfide, carbon dioxide, mercury, and other hydrocarbons. The application and enhancement of different operations including sweetening and dehydration make the raw natural gas ready for usage or conversion. The conversion of oil and gas is mainly conducted through refineries and chemical processing plants. Consequently, oil and gas refining leads through the combination of several physical, chemical, and thermodynamic processes to the formation of many valuable products including fuels, materials, chemicals, and/or heat and power. To note that even the polluting emissions of the refineries that can be formed at any stage of the oil or gas conversion process have a certain economic value and can be further processed or sold. Moreover, impurities removed during the oil and gas processing step are not always considered as waste, but rather potential by-products of that industry. The goal of refineries is generally to successfully implement an environmentally friendly process that provides an economic and environmental benefit.

This Special Issue on "Processing and Conversion of Oil and Gas: Modeling, Control, Simulation and Optimization" (https://www.mdpi.com/journal/processes/special_issues/oil_conversion, accessed on 21 January 2023) of *Processes* collects the recent work of leading researchers on modeling, control, simulation, and optimization of new and revamped petrochemical processes using proven software. Therefore, it offers novel illustrative examples, prospective applications, and solutions to improve these processes.

2. General Methods

Some papers employ general and practical methods of chemical process engineering through a holistic approach. The article by Al-Rabiah et al. [1] proposed the non-random two-liquid (NRTL) model in Aspen Plus to determine the vapor-liquid-liquid equilibrium by simulating the esterification reaction. This method was applied to design and model a membrane reactor-based process for an annual plant capacity of 92,500 metric tons of butyl acetate. Moreover, the UNIFAC method was applied as a thermodynamic model to simulate the pill particle discharge process test [2] and to develop a novel process for the production of methyl isobutyl ketone (MIBK) from hydrogen and acetone [3]. Lui et al. [2] studied the discharge process and its mechanism using the discrete element method (DEM) with self-developed annular corrosion pill particles and the discharge device to optimize the oil and gas field wellbore casing annular corrosion process. To simulate the pill particle discharge process simulation test, the EDEM software Hertz–Mindlin contact mechanics model was also employed. Additionally, Al Rabiah et al. [3] applied the UNIFAC method using CHEMCAD v.7.1 software to develop a process flow diagram for the production of methyl isobutyl ketone (MIBK) based on a production of 30,000 metric tons of MIBK per annum with 30 days assigned for maintenance. This study proposed a Generalized Additive Model based on 21 recognized dimensionless numbers to predict the pressure drop of gas–liquid two-phase flow based on 5011 records in the Tulsa Unified Fluid Flow Project (TUFFP) database. The process was heat integrated, resulting in a 26% and a 19.5% reduction in the heating and cooling utilities, respectively, leading to a 12.6% reduction in the total energy demand.

Cepeda-Vega et al. [4] proposed a Generalized Additive Model (GAM) to calculate the pressure drop in a gas–liquid two-phase flow at horizontal, vertical, and inclined pipes based on 21 different dimensionless numbers. The GAM non-parametric method achieved a high prediction capacity and permitted a good degree of interpretability describing each predictor's marginal effects, unlike in other machine Learning methods. A mean relative error of at most 19.98% for stratified flow and 12.93% for all the data points in a randomly sampled test set was obtained. Guo et al. [5] develop a mathematical model for the stability of a viscous liquid jet in a coaxial twisting compressible airflow. This model considers several factors such as the twist and compressibility of the surrounding airflow, the viscosity of the liquid jet, and the cavitation bubbles within the liquid jet, to be able to analyze the effects of aerodynamics caused by the gas–liquid velocity difference on the jet stability.

3. Processes Optimization

Several articles study certain processes to promote specialized models to improve their performance by examining and modifying the operating parameters. Therefore, the related studies will be described in the following. Oil quality improvement is investigated. Complex reaction extraction methods have been usually used to reduce nitrogen-containing compounds (NCs) from a coal tar fraction. Su Jin Kim [6] reduce NCs contained in wash oil by applying a batch equilibrium extraction. Wash oil and an aqueous solution of formamide is employed as the raw material and the solvent, respectively. The formamide extraction method is efficient with results to reduce the NCs of wash oil, and it is expected to be an alternative to the complex reaction extraction methods that have been applied. Al Rabiah et al. [7] present a promising novel industrial scheme for the synthesis of Dimethyl carbonate (DMC) via the oxidative carbonylation of vaporized methanol with dimethyl oxalate (DMO) as a byproduct. DMC and DMO are produced on a copper chloride catalyst in an isothermal FBR. A MeOH conversion rate of 81.86% and a DMC selectivity of 83.47% are reached. DMO purification is achieved through conventional distillation at 99.9 mol%. The pressure-swing technique is applied to separate the DMC-H_2O azeotropic mixture and a 99.78 mol% pure DMC product is obtained.

Encinar et al. [8] evaluate the usage of rapeseed oil as a starting point for a biorefinery. Biolubricant production through double transesterification is studied by modeling and designing the kinetics of the second transesterification and the reactor. A SAE 10W30 biolubricant suitable for Diesel engines, is obtained. A batch reactor, with a pseudo-first reaction order and a reactor volume of 9.66 m^3 is selected to produce this biolubricant in Spain.

Optimizing cleaning performance of reactors seems to be as important as process optimization. Therefore, Zhang et al. [9] studied the structure of self-excited oscillation pulsed jet nozzles (SOPJNs) to optimize cleaning and energy efficacies. The jet performance of a SOPJN was investigated and modeled based on computational fluid dynamics considering a large eddy simulation and homogeneous cavitation. The investigation includes several parameters such as inlet diameter, cavity diameter, cavity length, wall reflection angle, and inlet pressure on the jet's peak velocity, oscillation frequency, and cavitation number, cavity length, and wall reflection angle produced a jet with a high peak velocity and strong cavitation.

4. Processes Safety

Moreover, one article deals with pipelines risk. In fact, the shale gas collection and transportation pipeline causes significant risk due to certain geographical conditions and climatic conditions. Chen et al. [10] proposed a methodology that considers several failure scenarios including third-party damage, corrosion, design and construction defects, misoperation, and natural disasters. Since the used method employs subjective and objective data from different sources, an improved analytic hierarchy process was applied to process data and minimize subjectivity.

5. Economic Aspect

Finally, Ahmed et al. [11] analyzed the unexpected fuel price changes for the first time in Pakistan. The study aims to understand whether the fuel price driver is demand driven or whether it is exuberant consumer behavior that prevails and contributes to a sudden spike in the fuel price series. The empirical analysis is performed using a novel state-of-the-art generalized sup ADF (GSADF) approach on six commonly used fuel price series, such as LDO (light diesel oil), HSD (high-speed diesel), petrol, natural gas, kerosene, and MS (motor spirit).

6. Conclusions

To conclude, a broad variety of papers were presented. Many studies deal with general methods applicable to certain processes to improve the methods of process engineering. Other proposed studies are dedicated to improving and optimizing processes performance by examining and modifying the operating parameters, and improving the cleaning performance of reactors. One article analyzes safety risk of pipelines in China by studying several risk factors. Finally, the economic aspect was treated to understand the sudden fluctuation in fuel price in Pakistan.

Author Contributions: Writing—original draft preparation, J.C.A.; M.A.D.; writing—review and editing, J.C.A.; M.A.D. All authors have read and agreed to the published version of the manuscript.

Conflicts of Interest: The authors declare no conflict of interest.

References

1. Al-Rabiah, A.A.; Alqahtani, A.E.; Al Darwish, R.K.; Bin Naqyah, A.S. Novel Process for Butyl Acetate Production via Membrane Reactor: A Comparative Study with the Conventional and Reactive Distillation Processes. *Processes* **2022**, *10*, 1801. [CrossRef]
2. Liu, D.; Lu, Y.; Lin, H.; Qiao, C.; Song, J.; Chen, S.; Yao, Z.; Du, K.; Yu, Y. Study on the Discharge Process and Mechanism of Anti-Corrosion Pill Particles in the Oil and Gas Field Wellbore Casing Annulus Based on the Discrete Element Method. *Processes* **2022**, *10*, 1737. [CrossRef]
3. Al-Rabiah, A.A.; Alkathiri, R.R.; Bagabas, A.A. Process Development for Methyl Isobutyl Ketone Production Using the Low-Pressure One-Step Gas-Phase Selective Hydrogenation of Acetone. *Processes* **2022**, *10*, 1992. [CrossRef]
4. Cepeda-Vega, A.; Amaya-Gómez, R.; Asuaje, M.; Torres, C.; Valencia, C.; Ratkovich, N. Pipeline Two-Phase Flow Pressure Drop Algorithm for Multiple Inclinations. *Processes* **2022**, *10*, 1009. [CrossRef]
5. Guo, L.-M.; Lü, M.; Ning, Z. Stability of a Viscous Liquid Jet in a Coaxial Twisting Compressible Airflow. *Processes* **2021**, *9*, 918. [CrossRef]
6. Kim, S.J. Upgrading of Wash Oil through Reduction of Nitrogen-Containing Compounds. *Processes* **2021**, *9*, 1869. [CrossRef]

7. Al-Rabiah, A.A.; Almutlaq, A.M.; Bashth, O.S.; Alyasser, T.M.; Alshehri, F.A.; Alofai, M.S.; Alshehri, A.S. An Intensified Green Process for the Coproduction of DMC and DMO by the Oxidative Carbonylation of Methanol. *Processes* **2022**, *10*, 2094. [CrossRef]
8. Encinar, J.M.; Nogales-Delgado, S.; Pinilla, A. Biolubricant Production through Double Transesterification: Reactor Design for the Implementation of a Biorefinery Based on Rapeseed. *Processes* **2021**, *9*, 1224. [CrossRef]
9. Zhang, S.; Fu, B.; Sun, L. Investigation of the Jet Characteristics and Pulse Mechanism of Self-Excited Oscillating Pulsed Jet Nozzle. *Processes* **2021**, *9*, 1423. [CrossRef]
10. Chen, K.; Shi, N.; Lei, Z.; Chen, X.; Qin, W.; Wei, X.; Liu, S. Risk Classification of Shale Gas Gathering and Transportation Pipelines Running through High Consequence Areas. *Processes* **2022**, *10*, 923. [CrossRef]
11. Ahmed, M.; Irfan, M.; Meero, A.; Tariq, M.; Comite, U.; Abdul Rahman, A.A.; Sial, M.S.; Gunnlaugsson, S.B. Bubble Identification in the Emerging Economy Fuel Price Series: Evidence from Generalized Sup Augmented Dickey–Fuller Test. *Processes* **2022**, *10*, 65.

processes MDPI

Article

Novel Process for Butyl Acetate Production via Membrane Reactor: A Comparative Study with the Conventional and Reactive Distillation Processes

Abdulrahman A. Al-Rabiah *, Abdullah E. Alqahtani, Rayan K. Al Darwish and Abdulaziz S. Bin Naqyah

Chemical Engineering Department, College of Engineering, King Saud University, Riyadh 11421, Saudi Arabia
* Correspondence: arabiah@ksu.edu.sa; Tel.: +966-11-4676844; Fax: +966-11-4678770

Abstract: Butyl acetate (BuAc) is widely used as a solvent in many applications, mainly in the food and pharmaceutical industries. The conventional process for BuAc production is both capital and energy intensive. The purification process involves the separation of BuAc from the azeotropic mixture of water and *n*-butanol, which is difficult to accomplish using a simple distillation unit. In this study, a membrane reactor (MR) for BuAc production via the esterification of *n*-butanol was investigated. The MR using the Amberlyst-15 catalyst was modeled and validated with previously reported experimental data, and a good agreement was achieved. The ultimate conversion of *n*-butanol using the MR was 92.0%, compared to 69.8% for the conventional reactor. This study is the first to propose an intensified MR-based process for butyl acetate production. The MR-based process was developed and rigorously simulated using Aspen Plus for an annual plant capacity of 92,500 metric tons of BuAc. The MR-based process is environmentally friendly regarding CO_2 emissions, with a reduction of 80% compared to the conventional process. The economic analysis of the MR-based process shows a payback period of 2.7 years and a return on investment (ROI) of 23.1%. The MR-based process for BuAc production is a promising technology that provides similar key benefits as compared to the reactive distillation (RD) process.

Keywords: butyl acetate; acetic acid; *n*-butanol; esterification; membrane reactor; reactive distillation; amberlyst-15 catalyst; energy efficiency; economic evaluation

Citation: Al-Rabiah, A.A.; Alqahtani, A.E.; Al Darwish, R.K.; Bin Naqyah, A.S. Novel Process for Butyl Acetate Production via Membrane Reactor: A Comparative Study with the Conventional and Reactive Distillation Processes. *Processes* **2022**, *10*, 1801. https://doi.org/10.3390/pr10091801

Academic Editor: Jean-Claude Assaf

Received: 17 August 2022
Accepted: 5 September 2022
Published: 7 September 2022

1. Introduction

Butyl acetate (BuAc) is commonly used as a solvent across many industries, mostly in the manufacturing of adhesives, lacquers, coatings, and paints [1–6]. The pharmaceutical and cosmetic industries use butyl acetate as an extraction agent or solvent. In the food industry, BuAc is used to synthesize fruit flavors [7]. BuAc is produced on an industrial scale by *n*-butanol esterification with acetic acid. Strong acids are required to catalyze this reversible reaction using solid-acidic catalysts such as ion exchange resins [8–11].

The membrane reactor (MR) is a catalytic reactor that combines reaction and separation processes to increase the reaction yield when the thermodynamic equilibrium limits the conversion [12,13]. The separation inside the MR is performed through a selective membrane that allows some components to permeate through to the other side. In this way, the thermodynamic is shifted to an upper limit so the reaction can proceed forward to reach a higher conversion. Guangrui Liu et al. experimentally studied the synthesis of BuAc using the esterification of *n*-butanol through a catalytic membrane reactor (CMR) at 363 K [14]. Liu et al. studied the performance of pervaporation-assisted esterification using a cross-linked polyvinyl alcohol (PVA) membrane [15]. Khajavi et al. studied the esterification reaction in a hydroxy sodalite membrane reactor catalyzed by Amberlyst-15 [16]. Many studies have been published on testing different membrane reactors and water flux through different membranes under various reactive conditions. For instance, Zeolite A, Zeolite T, MOR, and NaA are water-selective membranes that were experimentally tested for the esterification of ethanol with lactic acid [17–19]. Although they were tested for different reactions, they showed good results for water flux and selectivity, which could be used for *n*-butanol esterification.

The conventional process for butyl acetate synthesis was studied by Shen et al., and a continuous stirred-tank reactor (CSTR) and three distillation columns were utilized for the process configuration [20]. The downside of the conventional process is that the esterification reaction is limited by the thermodynamic equilibrium. The conversion is typically low and does not exceed 70% [21]. In addition, the downstream purification process of BuAc using distillation columns is relatively complex and capital-intensive due to the azeotropic behavior of *n*-butanol and water [8].

Reactive distillation (RD) significantly reduces the energy and capital costs of the process by combining the reaction and separation equipment into one operating unit [22]. Several studies in the literature investigated the utilization of reactive distillation for the production of butyl acetate [2,8,23,24]. The RD technology can be utilized in a conventional multi-unit process to reduce energy and capital costs by five times [25]. However, one major drawback of RD technology is the complexity of the design and operation.

No previous work has been reported in the literature on developing an MR-based process for butyl acetate production via *n*-butanol esterification. Based on previous experimental studies, the goal of this research is to create a mathematical model to demonstrate the MR performance and validate the model results with the experimental data. The MR model is then utilized to develop a novel MR-based process for butyl acetate production. The new MR-based process is compared with the two well-known technologies: the conventional and the RD-based processes. The three processes are also evaluated based on technical performance, environmental impact, and economic profitability.

2. Reaction Chemical Equilibrium

The esterification reaction of *n*-butanol (BuOH) with acetic acid (AA) is shown in Equation (1):

$$\mathrm{BuOH} + \mathrm{AA} \rightleftharpoons \mathrm{BuAC} + \mathrm{H_2O} \qquad \Delta H_{298} = -29.312\ \mathrm{kJ/mol} \tag{1}$$

Due to the non-ideality of the system in the esterification reaction, the non-random two-liquid (NRTL) equation of state has been selected to determine the vapor-liquid-liquid equilibrium (VLLE) [23]. The NRTL model included in Aspen Plus was used to simulate the esterification reaction. The change of equilibrium conversion of *n*-butanol with the reaction temperature was determined using the equilibrium reactor as shown in Figure 1. It was noticed that the conversion decreases as the temperature increases, which confirms that the *n*-butanol esterification reaction is exothermic, as shown by Equation (1).

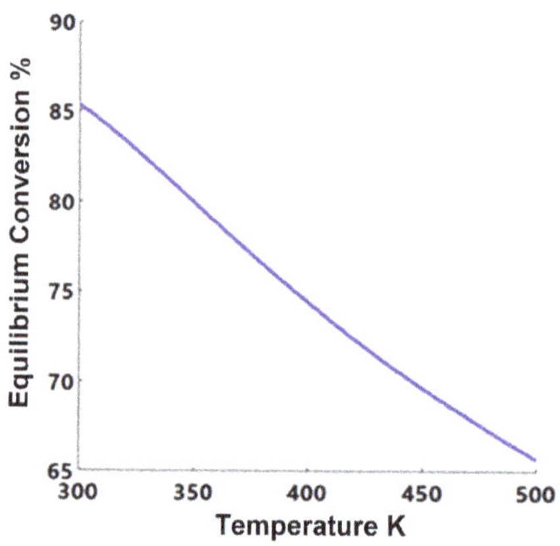

Figure 1. Equilibrium conversion of *n*-butanol vs. reaction temperature for the esterification reaction.

It should be emphasized that the n-butanol and water form an azeotropic mixture, which makes the separation more complicated. Shen et al. used 1,4-butanediol as a solvent to separate n-butanol from water [20]. A residue curve between n-butanol, water, and the solvent (1,4-butanediol) has been created using Aspen Plus, as shown in Figure 2. An azeotrope point is formed at 23.5% butanol and 76.5% water at a temperature of 91.6 °C (364.7 K).

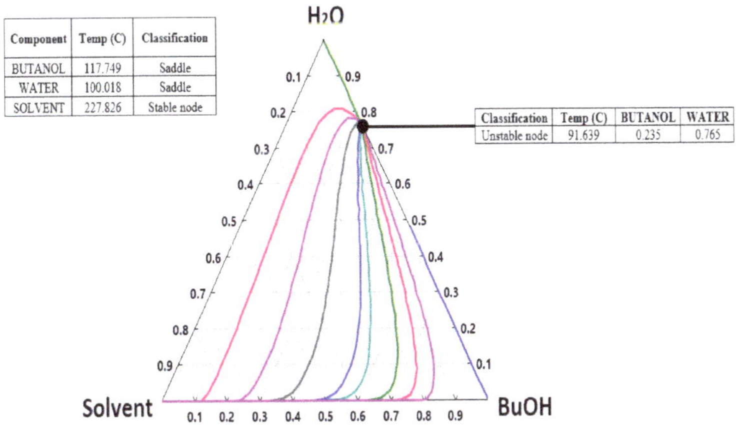

Figure 2. Residue curves map of the water-n-butanol-1,4-butanediol mixture at a pressure of 101.3 kPa.

3. Model Development

A mathematical model for the MR was developed based on the experimental data reported [16]. In the experimental work, a hydroxyl sodalite membrane was used in the reactor for the water separation from organic compounds.

The water permeation flux for the membrane reactor model was obtained from Khajavi et al. [16]. The Amberlyst-15 catalyst was used in the membrane reactor. The reaction rate expression using the Amberlyst-15 catalyst reported by Khajavi et al. is described in Equations (2)–(4) [16]:

$$rate\ of\ reaction\ (r) = k_F\left(C_{AA}C_{BuOH} - \frac{C_{BuAc}C_{H_2O}}{k_{eq}}\right) \tag{2}$$

$$k_F = 0.103\frac{\mathrm{L}}{\mathrm{mol\ min}} \tag{3}$$

$$k_{eq} = 3.51 \tag{4}$$

where r is the reaction rate, C_i is the concentration of the individual species, k_F is the reaction rate constant, and k_{eq} is the equilibrium constant.

The concentration of species i is expressed by:

$$C_i = N_i/V \tag{5}$$

where

$$V = \sum N_i\frac{M_i}{\rho_i} \tag{6}$$

N_i is the number of moles, V is the reaction volume, M_i is the molar mass, and ρ_i is the density.

The change in the number of moles over time of species i is expressed as:

$$\frac{dN_i}{dt} = -r_i V - J_i A \tag{7}$$

where r_i is the formation rate of component i, J_i is the membrane flux of component i, A is the effective area of the membrane.

Since there are only traces of n-butanol, butyl acetate, and acetic acid that can permeate through the membrane, these traces are assumed to be negligible. Therefore, the total flux can be assumed to equal the water flux (J_W), which is calculated by Equation (8).

$$J_W = P_W C_W \tag{8}$$

where P_W is the permeability coefficient of water.

Equations (9)–(12) were used to solve for the flowrates of the components along the length of the membrane reactor:

$$\frac{1}{A_c}\frac{d(F_{BuOH})}{d(z)} = -k_F\left(C_{AA}C_{BuOH} - \frac{C_{BuAc}C_{H_2O}}{k_{eq}}\right) - J_{BuOH}A \tag{9}$$

$$\frac{1}{A_c}\frac{d(F_{AA})}{d(z)} = -k_F\left(C_{AA}C_{BuOH} - \frac{C_{BuAc}C_{H_2O}}{k_{eq}}\right) - J_{AA}A \tag{10}$$

$$\frac{1}{A_c}\frac{d(F_{BuAc})}{d(z)} = k_F\left(C_{AA}C_{BuOH} - \frac{C_{BuAc}C_{H_2O}}{k_{eq}}\right) - J_{BuAc}A \tag{11}$$

$$\frac{1}{A_c}\frac{d(F_W)}{d(z)} = k_F\left(C_{AA}C_{BuOH} - \frac{C_{BuAc}C_{H_2O}}{k_{eq}}\right) - J_W A \tag{12}$$

where F_i is the flowrate of component i, A_c is the cross-sectional area, and z is the length of the membrane reactor. The model was solved using ODE45, which is based on an explicit Runge–Kutta (4,5) formula embedded in MATLAB software, R2020a (9.8) (Portola Valley, CA, USA).

4. Membrane Reactor Performance

The developed model was validated to accurately simulate the membrane reactor performance. A feed ratio of 1:1 for n-butanol and acetic acid was used in the experimental work. Table 1 lists the parameters used for the mathematical model [16].

Table 1. Membrane reactor model parameters.

MR Model Parameters	
BuOH flowrate	100 (kmol/h)
AA flowrate	100 (kmol/h)
Reactor length	8 m
Membrane effective surface area	100 m^2
MR cross sectional area	12.5 m^2
Permeability coefficient of water (P_W)	0.00148 (m^{-2}·h^{-1})

Figure 3 illustrates the concentration as a function of time obtained by the model and compares it with the experimental work conducted by Khavaji et al. [16]. The obtained results were in good agreement with the experimental work. Figure 4 illustrates the statistical variation of concentration results of theoretical calculations vs. experimental with the resultant coefficient of determination (R2) of 0.9973. The coefficient of determination indicates a good agreement between the experimental and theoretical results. Table 2 shows the model conversion results in comparison with the experimental data and their relative deviation.

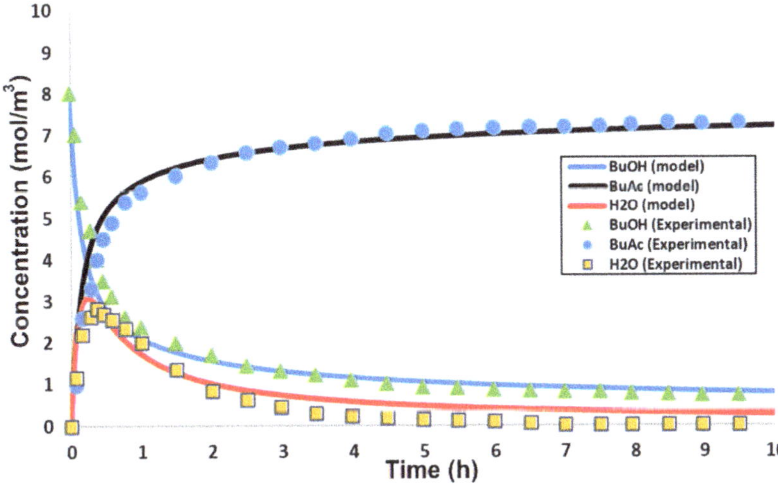

Figure 3. Components concentration as a function of the residence time of the membrane reactor.

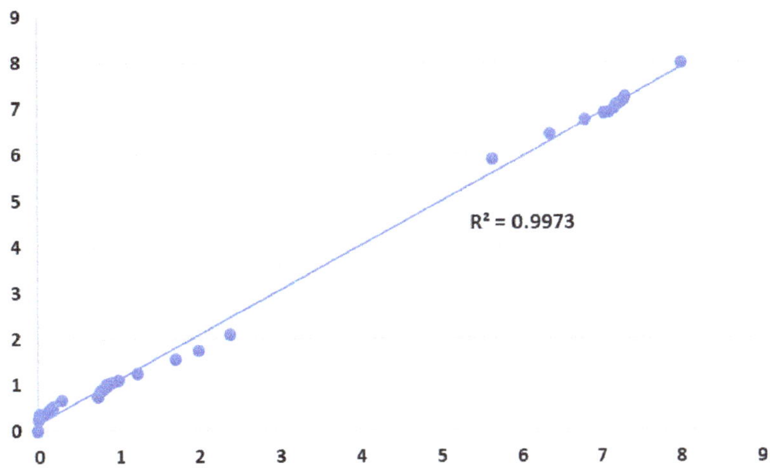

Figure 4. Variation of concentration results of mathematical model vs. experimental.

Table 2. Model results and experimental data using MR for BuAc production.

Sample	Conversion (%)		Relative Deviation (%)
	Experimental Results [16]	Model Results	
1	78.675	80.75	2.637
2	86.125	85.5	−0.725
3	89.125	87.875	−1.403
4	90.375	89.125	−1.383

Water is separated from the membrane reactor until it is almost removed as the permeate stream. Figure 5a shows the conversion of *n*-butanol in the membrane reactor as a function of MR residence time. The conversion achieves 92% at the end of the membrane reactor, which is similar to the previous experimental work [16].

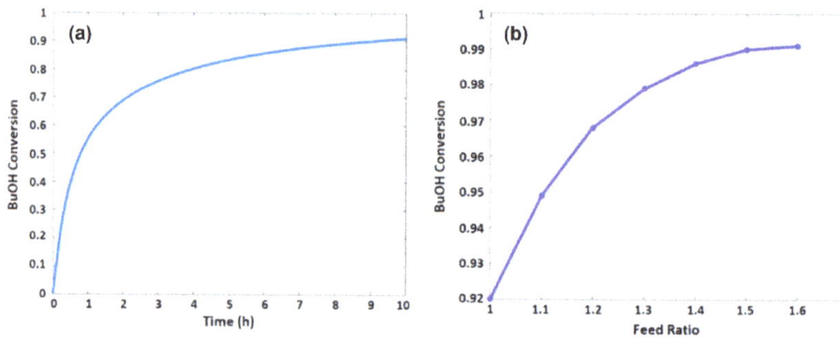

Figure 5. n-Butanol conversion as a function of (**a**) the residence time of the membrane reactor and (**b**) acetic acid/n-butanol feed ratio.

Figure 5b shows n-butanol conversion at different feed ratios of acetic acid to n-butanol. It is evident that the increase in acetic acid to n-butanol feed ratio increases the n-butanol conversion. However, the increase in the feed ratio will eventually cause more problems during product purification. The flowrates of reaction species and n-butanol conversion as a function of reactor length are shown in Figures 6 and 7, respectively.

The results obtained from the MR model were used in Aspen Plus to simulate the membrane reactor process, as described in the next section.

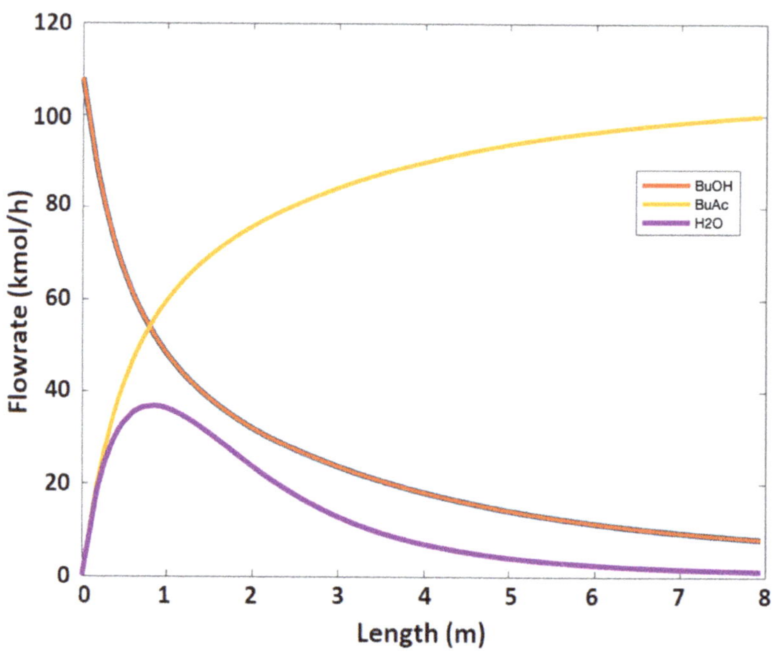

Figure 6. Components flow rates as a function of the length of the membrane reactor.

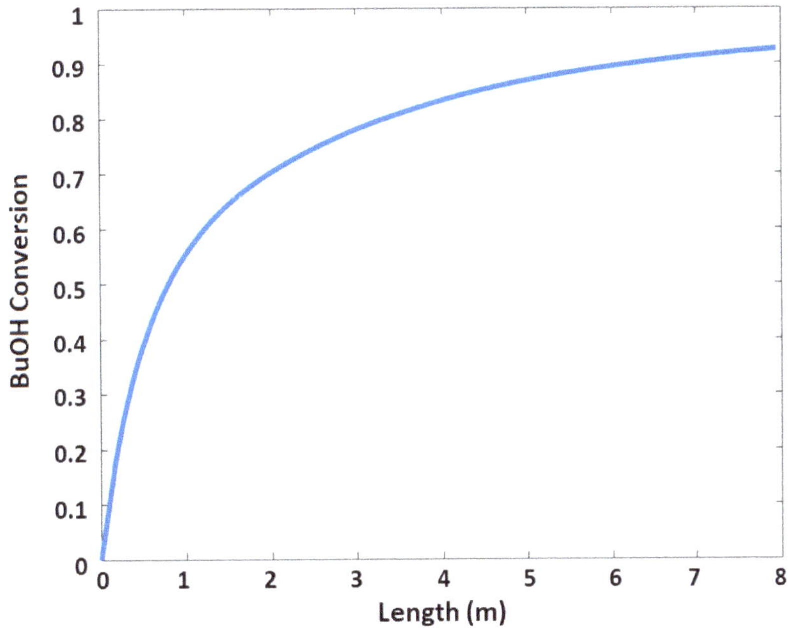

Figure 7. *n*-Butanol conversion as a function of the length of the membrane reactor.

5. Membrane Reactor-Based Process

The membrane reactor-based process for butyl acetate production was developed and simulated using Aspen Plus. Simulation software such as Aspen Plus has built-in models and databases that can be utilized to design, optimize, and control processes [23].

The developed MR-based process flowsheet is shown in Figure 8. A liquid *n*-butanol (stream 1) and a liquid acetic acid (stream 2) at a temperature of 298 K and a pressure of 111 kPa are mixed with the recycled acetic acid and *n*-butanol in the vessel, V-101. The feed stream (stream 3) is pumped into the membrane reactor (R-101) at a pressure of 304 kPa and a temperature of 313 K. The MR reactor is operated isothermally at a temperature of 363 K and a pressure of 280 kPa. The design parameters for the membrane reactor (R-101) are shown in Table 3. Butyl acetate and water are formed over the heterogeneous catalyst Amberlyst-15. As the reaction proceeds, water (stream W) is removed from the membrane reactor. Traces of other organic components dissolved in water have been considered. The MR effluent (stream 4) is pumped to a distillation column (T-101) for further purification. Figure 9a shows that the *n*-butanol and butyl acetate form an azeotropic mixture at atmospheric pressure. Figure 9b illustrates that the separation is possible at 304 kPa, but it requires a large number of stages. Figure 10 shows the liquid-phase molar compositions of *n*-butanol, acetic acid, and water inside the distillation column T-101 (Figure 8). The unreacted acetic acid and *n*-butanol are separated at the top of T-101 (stream 5) and recycled to the feed vessel, V-101. The product with a purity of 99.0 mol% (stream 6) leaves the bottom of the distillation column, and it is cooled down in heat exchanger E-101 to be sent to a storage tank (stream 6). The stream results of the process are shown in Table 4. Table 5 shows a summary of the major equipment parameters for the MR process.

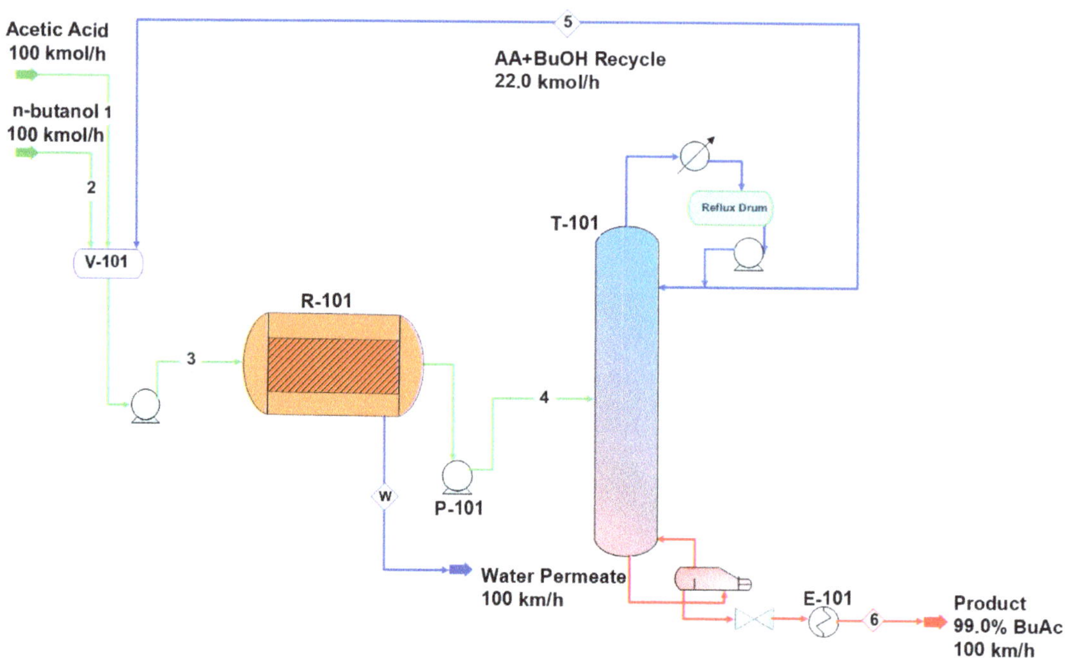

Figure 8. Process flow diagram (PFD) of butyl acetate production using a membrane reactor.

Table 3. Reactor design specifications for the membrane reactor process.

Membrane Reactor Specification	
BuOH to AA feed ratio	1:1
Temperature	363 K
Reactor pressure	284 kPa
Membrane surface area	100 m^2
Catalyst	Amberlyst-15

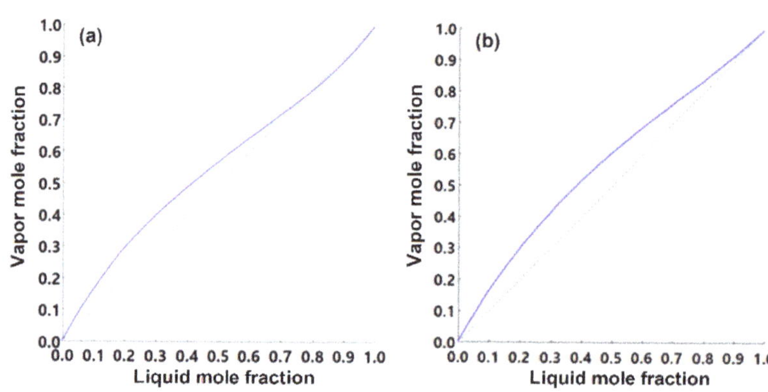

Figure 9. Equilibrium x-y diagram of n-butanol/butyl acetate mixture at: (a) 101 kPa and (b) 304 kPa.

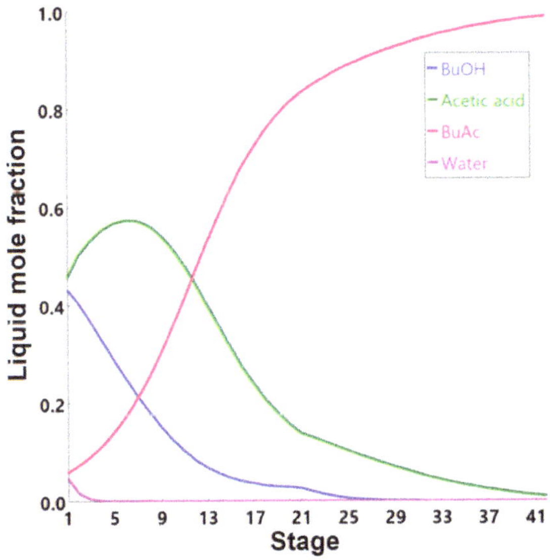

Figure 10. Liquid-phase molar compositions of components inside the extractive distillation column as a function of stage number.

Table 4. Stream information of the MR-based process for *n*-butanol esterification with acetic acid.

Stream	**1**	**2**	**3**	**4**
Temperature (K)	298.2	298.2	311	362.1
Pressure (kPa)	111	111	304	304
Enthalpy (kW)	−12,825	−9129	−24,042	−16,529
Molar vapor fraction	0	0	0	0
Mass flow (kg/h)	6065	7412	14,825	13,010
	Component Flowrates in (kmol/h)			
Butanol	0	100	108.69	8.69
Acetic acid	101	0	110.12	10.12
Butyl-acetate	0	0	1.18	101.07
Water	0	0	1.01	1.01
Stream	**5**	**6**	**7**	**W**
Temperature (K)	422.4	444.9	313.2	362
Pressure (kPa)	284	3.2	132	223
Enthalpy (kW)	−2090	−13,756	−14,727	−7818
Molar vapor fraction	0	0	0	0
Mass flow (kg/h)	1347	11,663	11,663	1814
	Component Flowrates in (kmol/h)			
Butanol	8.69	0	0	0.01
Acetic acid	9.12	1	1	0.01
Butyl-acetate	1.18	99.89	99.89	0.1
Water	1.01	0	0	99.99

Table 5. Major equipment parameters summary for the MR-based process.

Equipment	Parameter	Specifications
Membrane Reactor R-101	Volume	100 m^3
	Length	8 m
	Heat duty	−286 kW
	Temperature	363 K
	Pressure	223 kPa
	Acetic acid/n-butanol	1:1
Distillation column T-101	Number of stages	42 stages
	Feed stage	21
	Reboiler heat duty	5208 kW
	Condenser heat duty	−4526 kW
	Distillate rate	20 kmol/h
	Bottoms rate	100.9 kmol/h
Heat exchanger E-101	Heat duty	−970 kW
	Outlet temperature	313 K
	Outlet pressure	132 kPa
P-101	Discharge pressure	304 kPa
	Head	11 m
	Fluid power	0.92 kW
P-102	Discharge pressure	304 kPa
	Head	23 m
	Fluid power	0.36 kW

6. Alternative Processes

The alternative processes for butyl acetate synthesis are the conventional one, which is based on an isothermal catalytic fixed-bed reactor, and reactive distillation. Both processes are simulated in this study using Aspen Plus software for the same plant capacity and catalyst. The input data sets for each process were obtained from the literature [20,23]. All alternative processes were developed for equal annual plant capacity and the same product purity.

6.1. Conventional Process

The conventional process for the esterification of *n*-butanol using a fixed bed reactor is shown in Figure 11. The kinetic model used for the *n*-butanol esterification reaction is based on the commercial Amberlyst-15 catalyst and is represented by Equations (13)–(15) [1]:

$$rate\ of\ reaction\ (r) = k_F C_{AA} C_{BuOH} - k_B C_{BuAc} C_{H_2O} \tag{13}$$

$$k_F = \exp\left(11.472 - \frac{6986.3}{T}\right) \tag{14}$$

$$k_B = \exp\left(12.482 - \frac{7937.7}{T}\right) \tag{15}$$

where r is the reaction rate, C_i is the concentration, k_F, and k_B are the forward and backward reaction rate constants, respectively, and T is the reaction temperature (K). Figure 12 illustrates the component flowrate as a function of reactor length using the kinetic model equations, Equations (17)–(19).

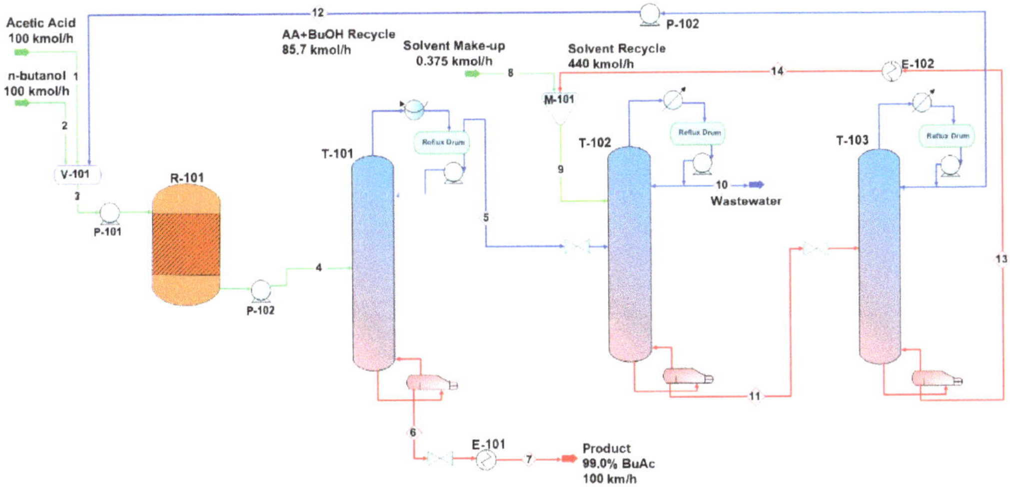

Figure 11. Conventional process flow diagram (PFD) for n-butanol esterification with acetic acid.

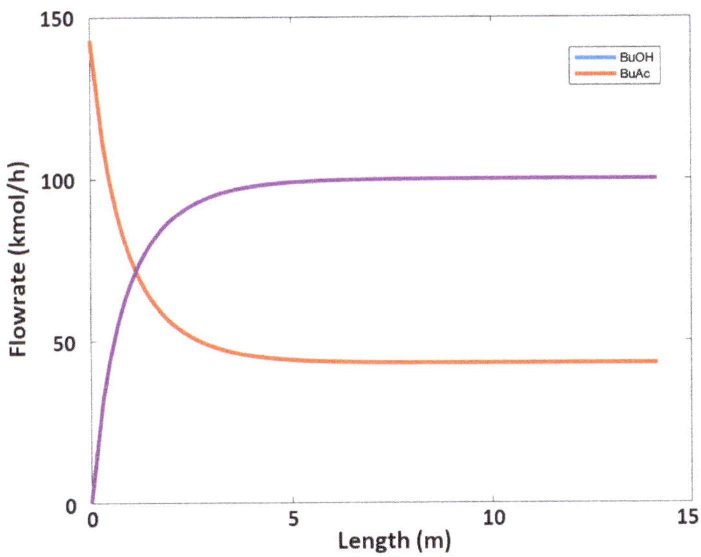

Figure 12. Components flowrates as a function of reactor length using the kinetic model.

The plant was designed to produce 92,500 metric tons of butyl acetate annually. Acetic acid (stream 1) and n-butanol (stream 2) are fed to the process and mixed in vessel V-101. A recycle (stream 12) that contains acetic acid and n-butanol is recycled to mix with the feed in V-101. The fresh feed and recycled stream are pumped by P-101 and enter a fixed-bed reactor, R-101, with a feed ratio of 1:1. Table 6 lists the design parameters and assumptions of reactor R-101. Figure 13 presents the equilibrium and kinetic conversions of n-butanol as a function of reaction temperature. The fixed-bed reactor process was simulated using the kinetic reactor (RPlug) of Aspen Plus. It was noticed that the maximum n-butanol conversion that could be achieved is ~70%. Figure 14 shows the n-butanol conversion as a function of reactor length, and the conversion increases exponentially up to 69.8% for a length of ~2.5 m. The reactor effluent is pumped with a pressure of 304 kPa using P-102 into a distillation column (T-101) to separate the butyl acetate from the other components. Butyl acetate leaves the bottom of the distillation column and is cooled in E-101. The unreacted acetic acid and *n*-butanol are separated with water at the top of the distillation column (T-101). This distillate stream is sent to an extractive distillation system for the separation of the azeotropic mixture, and 1,4-butanediol is used as a solvent in the extraction process at the top of the second column (T-102). Figure 15 shows the liquid-phase molar compositions of *n*-butanol, acetic acid, and water inside the extractive distillation column, which consists of 29 stages. Water leaves the extractive distillation process at the top of the column and is sent to a wastewater treatment unit. The solvent and unreacted reactants from the extractive distillation column (stream 11) are sent to a solvent recovery column (T-103), and the recovered solvent is recycled to the extractive distillation column (T-101). Acetic acid and *n*-butanol released from the top of the recovery column are pumped and recycled to the feed vessel (V-101). The stream information of the conventional process is shown in Table 7.

Table 6. Reactor design specifications for the conventional process.

Conventional Reactor Specifications	
BuOH-to-AA feed ratio	1:1
Temperature	363 K
Feed pressure	284 kPa
Volume	41 m^3
Catalyst	Amberlyst-15

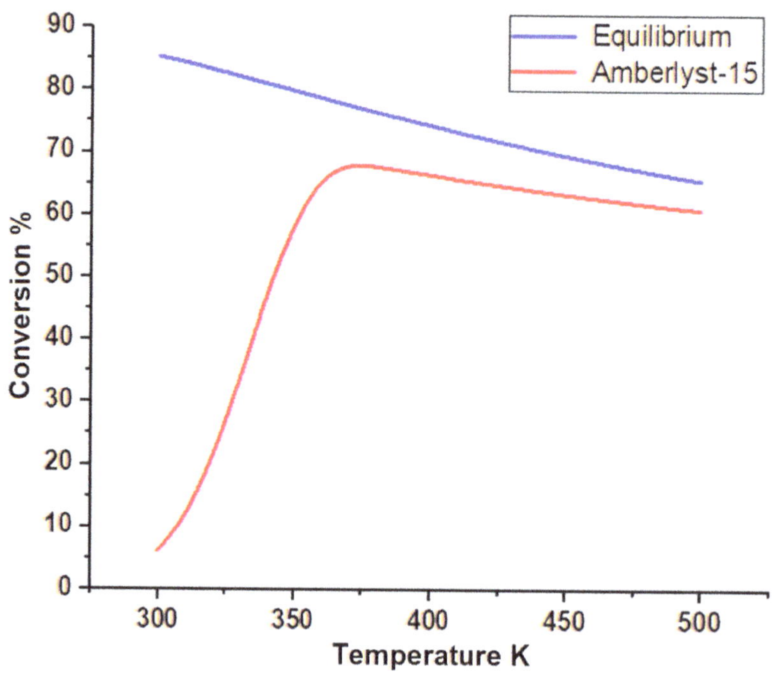

Figure 13. *n*-butanol conversion as a function of temperature for the esterification process.

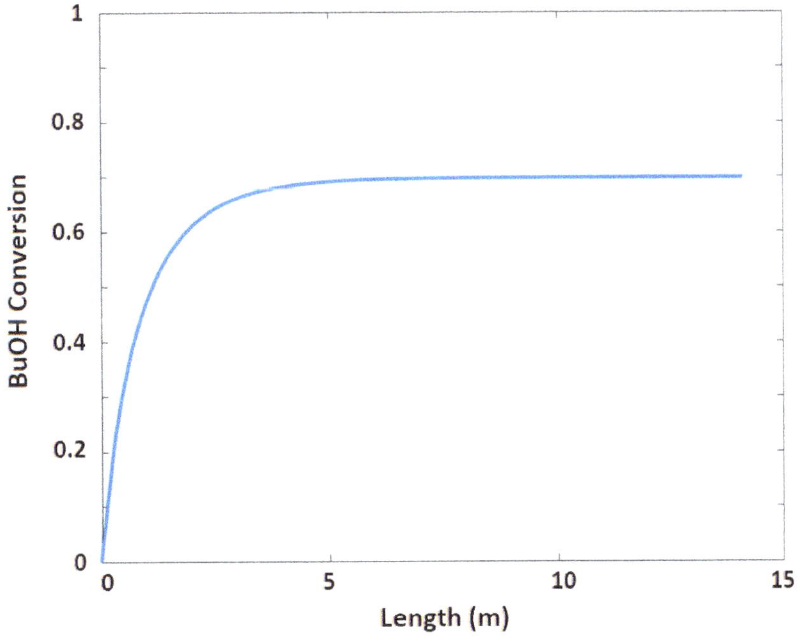

Figure 14. *n*-butanol conversion as a function of reactor length.

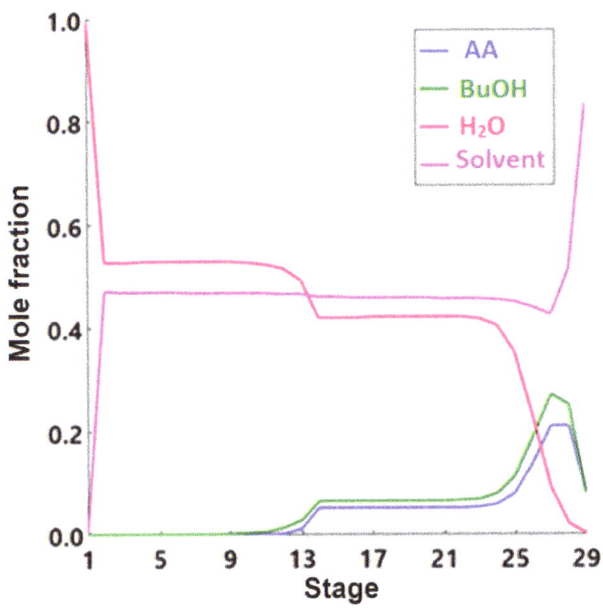

Figure 15. Liquid-phase molar compositions of components inside the extractive distillation column as a function of stage number.

Table 7. Streams information on the conventional process for *n*-butanol esterification with acetic acid.

Stream	**1**	**2**	**3**	**4**	**5**	**6**	**7**
Temperature (K)	298.2	298.2	298.2	363.3	411.1	444.9	308.2
Pressure (kPa)	111	111	111	304	284	324	132
Enthalpy (kW)	−18,158	−13,054	−31,212	−31,138	−14,805	−13,641	−14,639
Molar vapor fraction	0	0	0	0	1	0	0
Mass flow (kg/h)	6059	7417	19,278	19,278	7610	11,668	11,668
			Component Flowrates in (kmol/h)				
n-Butanol	0.0	100.1	143.1	43.2	43.1	0.1	0.1
Acetic acid	100.9	0.0	142.7	42.8	41.8	1.0	1.0
Butyl-acetate	0.0	0.0	0.9	100.8	0.9	99.9	99.9
Water	0.0	0.0	0.0	99.9	99.9	0.0	0.0
Solvent	0.0	0.0	0.0	0.0	0.0	0.0	0.0
Stream	8	9	10	11	12	13	14
Temperature (K)	298.2	507.0	373.1	462.6	393.6	512.2	303.2
Pressure (kPa)	121	121	111	142	132	142	132
Enthalpy (kW)	−54	−55,646	−7800	−65,897	−9060	−55,592	−61,735
Vapor mole fraction	0	0	0	0	0	0	0
Mass flow (kg/h)	34	39,687	1842	45,455	5802	39,653	39,653
			Component Flowrates in (kmol/h)				
n-Butanol	0.0	0.0	0.1	43.0	43.0	0.0	0.0
Acetic acid	0.0	0.0	0.0	41.8	41.8	0.0	0.0
Butyl-acetate	0.0	0.0	0.0	0.9	0.9	0.0	0.0
Water	0.0	0.0	99.9	0.0	0.0	0.0	0.0
Solvent	0.4	440.3	0.4	439.9	0.0	439.9	439.9

6.2. Reactive Distillation (RD) Process

The esterification of *n*-butanol with acetic acid is carried out in a reactive distillation, as demonstrated in Figure 16. The process consists of a reactive distillation column, a decanter, and three heat exchangers. The RD consists of 34 theoretical stages, which include a condenser and a kettle reboiler. The reactive zone starts at the 5th stage and ends at the 24th stage of the column. The acetic acid and *n*-butanol are fed in a 1:1 ratio, and the achieved conversion is more than 98% in the process. The acetic acid (stream 1) with a flowrate of 100 kmol/h is fed to the RD at the fifth stage, while the *n*-butanol at a 100 kmol/h flowrate (stream 2) is fed to the RD at the ninth stage. Table 8 lists the design parameters for the RD process.

Table 8. Reactor design specifications for the RD process.

Reactive Distillation Design Specifications	
Number of stages	34
AA feed stage	5
BuOH feed stage	9
Reactive zone stages	5–24
Reboiler duty	3298 kW
Condenser duty	−3010 kW

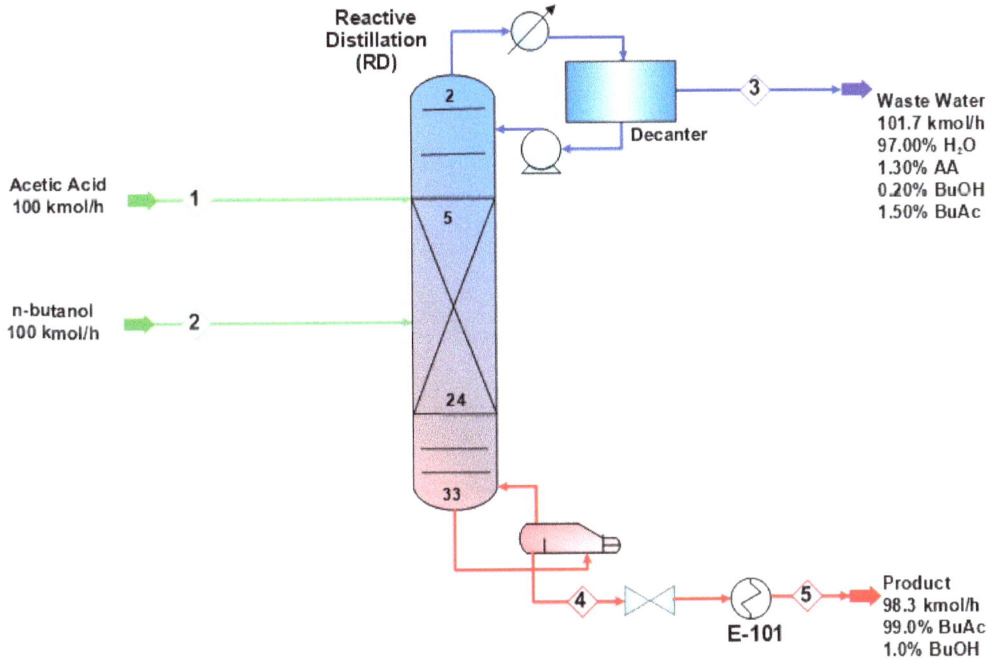

Figure 16. *n*-Butanol esterification with acetic acid by reactive distillation process.

A decanter is used to separate the distillate into the water and organic phases. Wastewater of 101.6 kmol/h (stream 3) is sent to a wastewater treatment unit while the organic phase of the decanter is sent back to the column. The butyl acetate product is separated at the bottom of the RD with a purity of more than 99%. The product is sent to a cooler (E-101) and reduced to a temperature of 313 K. The stream information of the reactive-based process is shown in Table 9. The liquid-phase compositions and the temperature profile inside the reactive distillation column are shown in Figure 17a,b, respectively.

Table 9. Stream information of the reactive distillation-based process for *n*-butanol esterification with acetic acid.

Stream	1	2	3	4	5
Temperature (K)	298.2	298.2	382.6	428.5	313.2
Pressure (kPa)	203	203	152	223	132
Enthalpy (kW)	−12,698	−9129	−8045	−13,493	−14,312
Molar vapor fraction	0	0	0	0	0
Mass flow (kg/h)	6005	7412	2040	11,377	11,377
Component Flowrates in (kmol/h)					
n-Butanol	0.0	100.0	0.2	1.1	1.1
Acetic acid	100.0	0.0	1.3	0.0	0.0
Butyl-acetate	0.0	0.0	1.5	97.2	97.2
Water	0.0	0.0	98.7	0.0	0.0

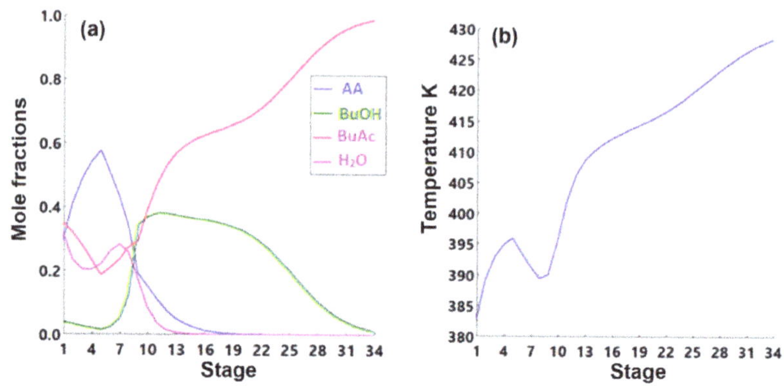

Figure 17. (**a**) The liquid-phase molar compositions of components as a function of stage number. (**b**) Temperature profile inside the reactive distillation column as a function of stage number.

7. Energy Efficiency and Environmental Analysis

7.1. Energy Efficiency Analysis

An energy efficiency analysis was conducted on MR, RD, and conventional processes. The steam consumption for the butyl acetate production by the MR, RD, and conventional processes was calculated based on the process utility requirements using Aspen Plus. The utilities used for the conventional, MR, and RD processes are cooling water at 293 K, low-pressure steam at 398 K, and medium-pressure steam at 448 K. Table 10 shows the utility requirements for the conventional, RD, and MR process configurations.

Table 10. Required hot and cold utilities in the three esterification processes.

Utility (MWh/Year)	Conventional	RD	MR
Steam (mps)	218,729	26,388	41,668
Cooling water	215,504	30,638	43,973

Based on the utility requirements of the three processes, it is shown that the conventional process requires more utility than the RD and MR processes. This is due to the requirements of the distillation columns in the conventional process. The MR process consumes less energy than the conventional process since it requires fewer distillation columns for product purification. However, since the RD distillation process has one RD distillation column, it requires lower energy consumption compared to the MR process. The heat duty for the MR and RD processes is 5208 kW and 3298 kW, respectively.

7.2. Environmental Analysis

An environmental assessment was conducted on the three distinct butyl acetate processes. The impact of the butyl acetate processes on the environment was evaluated based on CO_2 emissions. It is assumed that steam is generated using natural gas, and the CO_2 emitted from the butyl acetate process is mainly produced by the process of steam consumption. The Aspen Plus Environmental Analyzer was used to calculate the amount of CO_2 emitted from the three distinctive processes. The Aspen software determines the amount of generated CO_2 based on the fuel type. In this study, natural gas (NG) was used for all processes as a source of energy.

Figure 18 shows the emitted CO_2 per butyl acetate produced, and the conventional process shows the highest CO_2 emission compared to the MR and RD processes. This is due to the high steam required by the conventional process.

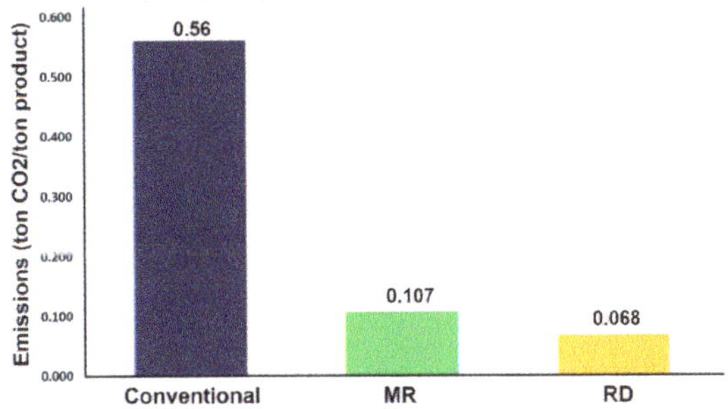

Figure 18. CO_2 emissions per product for the three processes.

8. Economic Analysis

An economic analysis was conducted for the membrane reactor, conventional and reactive distillation processes. The purchased equipment costs were determined using the Aspen Plus Economic Analyzer, and the Chemical Engineering Plant Cost Index (CEPI) of 2021 was used to update the equipment costs. Table 11 shows the prices of acetic acid, *n*-butanol, and butyl acetate, along with the prices of other utilities used for the process.

Table 11. Raw materials, products, and utility costs for the esterification process [23,26].

Unit	Price
Acetic Acid (USD/ton)	960
n-Butanol (USD/ton)	1610
Butyl Acetate (USD/ton)	2120
Utilities	
Steam (mps) (USD/ton)	6
Cooling water (USD/m^3)	0.53
Wastewater treatment (USD/m^3)	0.08

Figure 19 shows a comparison of direct costs, indirect costs, working capital, and total capital investment for the three processes. It was found that the total capital investment of the membrane reactor and reactive distillation processes is much lower compared to the conventional process. The separation of unreacted *n*-butanol and butyl acetate in the MR-based process is difficult and requires a distillation column with a larger number of stages. This ultimately increased the capital investment and operating costs compared to the RD process.

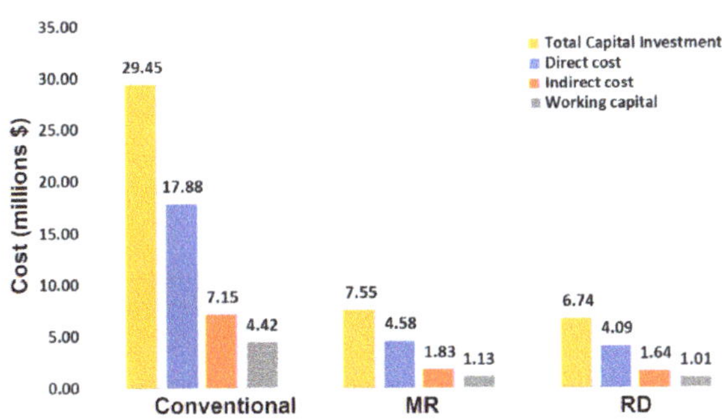

Figure 19. Economic comparisons of the three processes for *n*-butanol esterification with acetic acid.

Table 12 shows an economic comparison between the MR and RD processes for butyl acetate production. The economic indicators show that the MR process has a return on investment (ROI) of 23.1% and a payback period of 2.7 years. However, the MR process is economically less attractive compared to the RD process because of the capital and energy-intensive distillation column used for the separation of *n*-butanol and acetic acid from the product.

Table 12. Economic comparisons between the MR process and RD process for butyl acetate production.

Item	MR	RD
Fixed capital cost	USD 6,418,000	USD 5,730,000
Total capital cost	USD 7,552,000	USD 6,742,000
Return on investment	23.1%	31.17%
Payback period	2.7	2.1
Net present value	USD 13,670,000	USD 15,857,000

Figure 20 shows the cost distribution of the total production costs (TPC) for the MR-based process. The raw materials are responsible for around 87% of the TPC. Figure 21a,b show that the utility cost distributions of the MR and RD processes are similar.

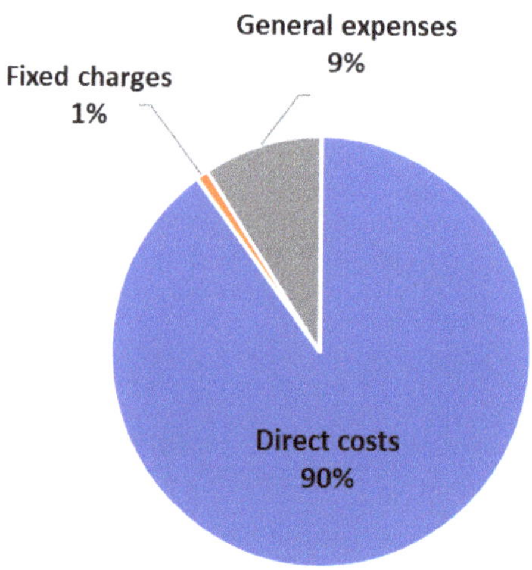

Figure 20. Cost distribution of the total production costs (TPC) of the process MR-based process.

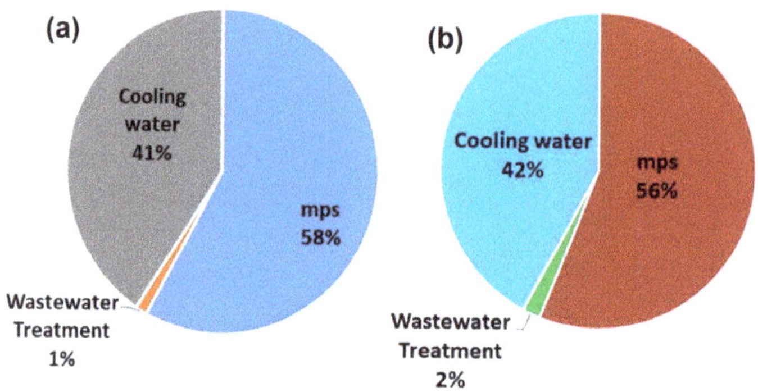

Figure 21. Annual utilities cost distribution of (**a**) MR-based process and (**b**) RD-based process.

9. Conclusions

A novel process based on a membrane reactor was designed for butyl acetate production. The membrane reactor was modeled and validated with the experimental data, and a good agreement was found. The MR-based process was developed and designed for an annual capacity of 92,500 metric tons of BuAc. The *n*-butanol conversion of MR, conventional, and RD processes was 92%, 69.8%, and 98%, respectively. The MR-based process is promising, economically feasible, and has less CO_2 emissions than the conventional process. However, the MR-based process is slightly less profitable when compared to the RD process due to the intensive separation that follows the membrane reactor. If the conversion of *n*-butanol in the membrane reactor increases to more than 95%, the MR process will be more attractive than the RD process. Future research should focus on finding more selective catalysts that can be used with the membrane for butyl acetate production.

Author Contributions: Conceptualization, A.A.A.-R.; methodology, A.A.A.-R. and A.E.A.; software, A.A.A.-R., A.E.A. and R.K.A.D.; validation, A.E.A., R.K.A.D. and A.A.A.-R.; formal analysis, A.A.A.-R., A.E.A., R.K.A.D. and A.S.B.N.; investigation, A.A.A.-R., A.E.A., R.K.A.D. and A.S.B.N.; resources, A.E.A. and A.A.A.-R.; data curation, A.E.A. and A.A.A.-R.; writing—original draft preparation, A.A.A.-R. and A.E.A.; writing—review and editing, A.A.A.-R. and A.E.A.; supervision, A.A.A.-R.; project administration, A.A.A.-R.; funding acquisition, A.A.A.-R. All authors have read and agreed to the published version of the manuscript.

Funding: This project was supported by King Saud University, Deanship of Scientific Research, College of Engineering Research Center.

Institutional Review Board Statement: Not applicable.

Informed Consent Statement: Not applicable.

Data Availability Statement: Not applicable.

Conflicts of Interest: The authors declare no conflict of interest.

References

1. Arpornwichanop, A.; Koomsup, K.; Assabumrungrat, S. Hybrid Reactive Distillation Systems for N-Butyl Acetate Production from Dilute Acetic Acid. *J. Ind. Eng. Chem.* **2008**, *14*, 796–803. [CrossRef]
2. Steinigeweg, S.; Gmehling, J. N-Butyl Acetate Synthesis via Reactive Distillation: Thermodynamic Aspects, Reaction Kinetics, Pilot-Plant Experiments, and Simulation Studies. *Ind. Eng. Chem. Res.* **2002**, *41*, 5483–5490. [CrossRef]
3. Tian, H.; Zhao, S.; Zheng, H.; Huang, Z. Optimization of Coproduction of Ethyl Acetate and N-Butyl Acetate by Reactive Distillation. *Chin. J. Chem. Eng.* **2015**, *23*, 667–674. [CrossRef]
4. Sato, T.; Nagasawa, H.; Kanezashi, M.; Tsuru, T. Enhanced Production of Butyl Acetate via Methanol-Extracting Transesterification Membrane Reactors Using Organosilica Membrane: Experiment and Modeling. *Chem. Eng. J.* **2022**, *429*, 132188. [CrossRef]
5. Fang, D.; Wen, Z.; Lu, M.; Li, A.; Ma, Y.; Tao, Y.; Jin, M. Metabolic and Process Engineering of Clostridium Beijerinckii for Butyl Acetate Production in One Step. *J. Agric. Food Chem.* **2020**, *68*, 9475–9487. [CrossRef]
6. Cao, X.; Ren, J.; Xu, C.; Zhang, K.; Zhan, C.; Lan, J. Preparation, Characterization of Dawson-Type Heteropoly Acid Cerium (III) Salt and Its Catalytic Performance on the Synthesis of n-Butyl Acetate. *Chin. J. Chem. Eng.* **2013**, *21*, 500–506. [CrossRef]
7. Ben Salah, R.; Ghamghui, H.; Miled, N.; Mejdoub, H.; Gargouri, Y. Production of Butyl Acetate Ester by Lipase from Novel Strain of Rhizopus Oryzae. *J. Biosci. Bioeng.* **2007**, *103*, 368–372. [CrossRef] [PubMed]
8. Hanika, J.; Kolena, J.; Smejkal, Q. Butylacetate via Reactive Distillation—Modelling and Experiment. *Chem. Eng. Sci.* **1999**, *54*, 5205–5209. [CrossRef]
9. Peters, T.A.; Benes, N.E.; Holmen, A.; Keurentjes, J.T.F. Comparison of Commercial Solid Acid Catalysts for the Esterification of Acetic Acid with Butanol. *Appl. Catal. A Gen.* **2006**, *297*, 182–188. [CrossRef]
10. Blagov, S.; Parada, S.; Bailer, O.; Moritz, P.; Lam, D.; Weinand, R.; Hasse, H. Influence of Ion-Exchange Resin Catalysts on Side Reactions of the Esterification of n-Butanol with Acetic Acid. *Chem. Eng. Sci.* **2006**, *61*, 753–765. [CrossRef]
11. Gangadwala, J.; Mankar, S.; Mahajani, S.; Kienle, A.; Stein, E. Esterification of Acetic Acid with Butanol in the Presence of Ion-Exchange Resins as Catalysts. *Ind. Eng. Chem. Res.* **2003**, *42*, 2146–2155. [CrossRef]
12. Abashar, M.E.E.; Al-Rabiah, A.A. Production of Ethylene and Cyclohexane in a Catalytic Membrane Reactor. *Chem. Eng. Process. Process Intensif.* **2005**, *44*, 1188–1196. [CrossRef]
13. Bin Naqyah, A.S.; Al-Rabiah, A.A. Development and Intensification of the Ethylene Process Utilizing a Catalytic Membrane Reactor. *ACS Omega* **2022**, *7*, 28445–28458. [CrossRef]

14. Liu, G.; Guo, S.; He, B.; Li, J.; Qian, X. Synthesis of Butyl Acetate in a Membrane Reactor in a Flow-Through Mode. *Int. J. Chem. React. Eng.* **2016**, *14*, 579–585. [CrossRef]
15. Liu, Q.; Zhang, Z.; Chen, H. Study on the Coupling of Esterification with Pervaporation. *J. Memb. Sci.* **2001**, *182*, 173–181. [CrossRef]
16. Khajavi, S.; Jansen, J.C.; Kapteijn, F. Application of a Sodalite Membrane Reactor in Esterification—Coupling Reaction and Separation. *Catal. Today* **2010**, *156*, 132–139. [CrossRef]
17. de la Iglesia, Ó.; Mallada, R.; Menéndez, M.; Coronas, J. Continuous Zeolite Membrane Reactor for Esterification of Ethanol and Acetic Acid. *Chem. Eng. J.* **2007**, *131*, 35–39. [CrossRef]
18. Tanaka, K.; Yoshikawa, R.; Ying, C.; Kita, H.; Okamoto, K. Application of Zeolite Membranes to Esterification Reactions. *Catal. Today* **2001**, *67*, 121–125. [CrossRef]
19. Jafar, J.J.; Budd, P.M.; Hughes, R. Enhancement of Esterification Reaction Yield Using Zeolite A Vapour Permeation Membrane. *J. Memb. Sci.* **2002**, *199*, 117–123. [CrossRef]
20. Shen, Y.; Zhao, F.; Qiu, X.; Zhang, H.; Yao, D.; Wang, S.; Zhu, Z.; Yang, J.; Cui, P.; Wang, Y.; et al. Economic, Thermodynamic, and Environmental Analysis and Comparison of the Synthesis Process of Butyl Acetate. *Ind. Eng. Chem. Res.* **2020**, *59*, 21869–21881. [CrossRef]
21. Liu, K.; Tong, Z.; Liu, L.; Feng, X. Separation of Organic Compounds from Water by Pervaporation in the Production of N-Butyl Acetate via Esterification by Reactive Distillation. *J. Memb. Sci.* **2005**, *256*, 193–201. [CrossRef]
22. Tian, H.; Huang, Z.; Qiu, T.; Wang, X.; Wu, Y. Reactive Distillation for Producing N-Butyl Acetate: Experiment and Simulation. *Chin. J. Chem. Eng.* **2012**, *20*, 980–987. [CrossRef]
23. Luyben, W.L.; Yu, C.C. *Reactive Distillation Design and Control*; John Wiley & Sons: Hoboken, NJ, USA, 2009.
24. Gangadwala, J.; Kienle, A.; Stein, E.; Mahajani, S. Production of Butyl Acetate by Catalytic Distillation: Process Design Studies. *Ind. Eng. Chem. Res.* **2004**, *43*, 136–143. [CrossRef]
25. Taylor, R.; Krishna, R. Modelling Reactive Distillation. *Chem. Eng. Sci.* **2000**, *55*, 5183–5229. [CrossRef]
26. Chemanalyst. 2022. Available online: https://www.chemanalyst.com/ (accessed on 2 May 2022).

processes

Article

Study on the Discharge Process and Mechanism of Anti-Corrosion Pill Particles in the Oil and Gas Field Wellbore Casing Annulus Based on the Discrete Element Method

Dongtao Liu [1,*], Yuliang Lu [1], Haichun Lin [2], Chunshang Qiao [1], Jiming Song [1], Shengqian Chen [1], Zhenhe Yao [1], Kezheng Du [1] and Yajun Yu [3,*]

1 CNOOC EnerTech-Drilling &Production Co., Shenzhen 518067, China
2 Shenzhen Branch, CNOOC(China) Co., Ltd., Shenzhen 518067, China
3 School of Biological and Agricultural Engineering, Jilin University, Changchun 130022, China
* Correspondence: liudt@cnooc.com.cn (D.L.); yuyajun@jlu.edu.cn (Y.Y.); Tel.: +86-138-2879-7030; +86-151-4317-3701

Abstract: This research studies the discharge process and its mechanism using the discrete element method (DEM) with self-developed annular corrosion pill particles and the discharge device as an example in order to optimize the oil and gas field wellbore casing annular corrosion process. The object of study was chosen from four different grid numbers and four different grid widths, and EDEM software was utilized to simulate and assess the pill particle discharge process based on preliminary experimental research. Under five different pill wheel rotation speeds, the effects of the grid number and grid width on the filling amount, filling density, discharge variation coefficient, and compressive force of pill particles were investigated from macroscopic and microscopic viewpoints. The findings reveal that the grid number, grid width, and rotation speed all have a significant impact on pill filling and discharge performance. As a result, the discharge wheel's structure and operating characteristics were optimized. The discharge wheel performs best when the grid number is 8, the grid width is 75 mm, and the rotation speed is 15 rpm; the pill filling density is 692.26 kg/m^3, the discharge variation coefficient is 0.022, and the maximum compressive force is 188 N. This study establishes the groundwork for enhancing wellbore integrity management in oil and gas fields by providing a reference for the optimal design of wellbore casing annular corrosion prevention devices in oil and gas fields.

Keywords: discrete element method; optimal design; simulation analysis; anti-corrosion pill particles

Citation: Liu, D.; Lu, Y.; Lin, H.; Qiao, C.; Song, J.; Chen, S.; Yao, Z.; Du, K.; Yu, Y. Study on the Discharge Process and Mechanism of Anti-Corrosion Pill Particles in the Oil and Gas Field Wellbore Casing Annulus Based on the Discrete Element Method. *Processes* **2022**, *10*, 1737. https://doi.org/10.3390/pr10091737

Academic Editors: Jean-Claude Assaf and Blaž Likozar

Received: 24 June 2022
Accepted: 30 August 2022
Published: 1 September 2022

1. Introduction

The conventional oil and gas field wellbore casing annulus corrosion prevention fluid filling procedure, while necessary for wellbore integrity maintenance, has major drawbacks, including a time-consuming construction process and a high cost [1–3]. As a result, our business produced a solid corrosion-inhibiting and anti-corrosion pill on its own, with the goal of developing a novel procedure for protecting the casing in offshore oil and gas wells against corrosion [4]. However, due to a number of constraints, such as the pill's fragility and small casing, the creation of a high-performance discharge mechanism that allows pill particles to enter the casing annulus in a uniform and stable manner has become a critical technical bottleneck that must be addressed.

The influence law of the fundamental design parameters of the pill wheel on the discharge process becomes a precondition and is crucial to improving the discharge device's functioning quality [5,6]. The grid number, grid width, and rotation speed of the discharge wheel are the main factors that govern the kinematics and dynamics of the pill particles throughout the discharge process [7–10]. As a result, in order to increase the pill discharge device's operation quality, the ideal combination of the pill discharge wheel's design parameter values must be improved.

The main issue in the optimal design of the parameters of the pill discharge wheel is obtaining precise observation and analysis of the kinematic and kinetic states of the pill particle population during the discharge operation [11]. Empirical and experimental procedures are currently the most often utilized research methods [12–15]. However, the approaches described above rely on a significant number of experiments, which are both time- and money-consuming and not general. In recent years, researchers have begun to investigate the use of DEM in the industrial and agricultural production domains for high precision numerical simulation of particle population dynamics states [16–18]. Most numerical simulations based on DEM to achieve accurate emission processes of particulate materials are now focused on agricultural products such as seeds and fertilizers [19–23]. However, the shape, size, and composition of the anti-corrosion pill particles analyzed in this work differ greatly from agricultural materials. As a result, the dynamics of the pill particle population and the high-precision simulation of the pill discharge device's working process must be thoroughly investigated.

In summary, the Hertz–Mindlin contact mechanics model was used to simulate the contact mechanics behavior between the pill particles and the discharge device in this research using EDEM simulation software. The flow process of the pill particle population was simulated using the well-established DEM numerical model of pill particles. The impacts of the structural and operating parameters of the discharge wheel on the performance of the pill filling amount, filling density, discharge variation coefficient, and compressive force of the pill particles were disclosed. Finally, the discharge wheel's structure and operating characteristics were optimized.

2. Models and Methods

2.1. Model of Pill Discharge Device

This paper independently designed a wheel-type pill discharge device, which included the main working parts, such as the material box, pill wheel, motor, frame, and frame and fluid conveyer, as shown in Figure 1a, in order to make the pill particles evenly and stably placed into the casing annulus of an oil and gas field wellbore.

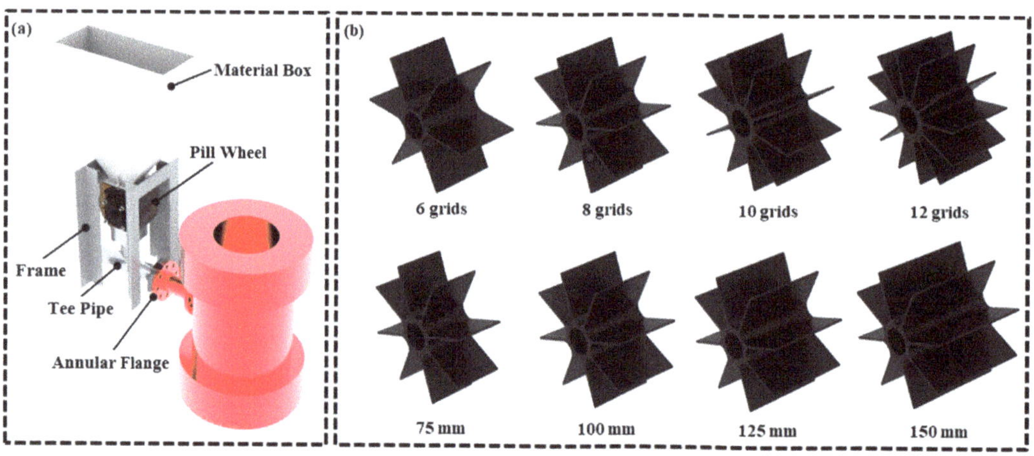

Figure 1. Entity diagram of the pill discharge device and different pill discharge wheels: (**a**) the pill discharge device; (**b**) different grid numbers and widths of pill discharge wheels.

The material box used galvanized steel as the contact material to increase the stability of the pill particle flow process and reduce adhesion between the pill powder and the material box, and the half angle of the conical top was designed according to the friction characteristics of the pill particles. The discharge wheel was composed of ABS plastic as the contact material and consisted of a set number of wheel grids to promote uniformity and prevent pill particle breakage during the filling process. The wheel drive shaft was composed of steel and was tightly attached to the frame and motor to ensure that the mechanical strength of the mechanism drive process did not fluctuate. The upper discharge pills were transported to the annular flange of the oil field gas well through a tee pipe using fluid conveying in order to minimize pill particle damage during the conveying operation. The behavior of fluid transportation and particle settlement were significantly influenced by the discharge procedure and performance. In order to obtain the ideal working state, this research focused primarily on the discharge device, its functioning, and its influencing mechanisms.

The lower arc of the trough wheel grid was tangential to the inner circle to maximize particle filling efficiency. As the discharge wheel revolved, the granules filled the grid uniformly and were discharged steadily at the outlet. The number and width of the grids have a big influence on particle filling efficiency and discharge stability. As a result, four grid numbers (6, 8, 10, and 12) and four grid widths (75, 100, 125, and 150 mm) were chosen for the study, as shown in Figure 1b, to evaluate the influence of the discharge wheel's shape and dimensional parameters on the discharge process' performance.

2.2. Model of Pill Particle

The research object in this work was self-developed pill particles. Depending on the geometrical features of the particles, the particles were divided into three groups based on their height, namely 5.4, 5.8, and 6.2 mm. In the particle population, the mass ratios of pill particles A, B, and C were 35%, 35%, and 30%, respectively. As a consequence, a strategy for modeling the population of pill particles based on DEM is presented, as well as numerical models of the three pill particles based on the 12-sphere model [4], as illustrated in Figure 2a.

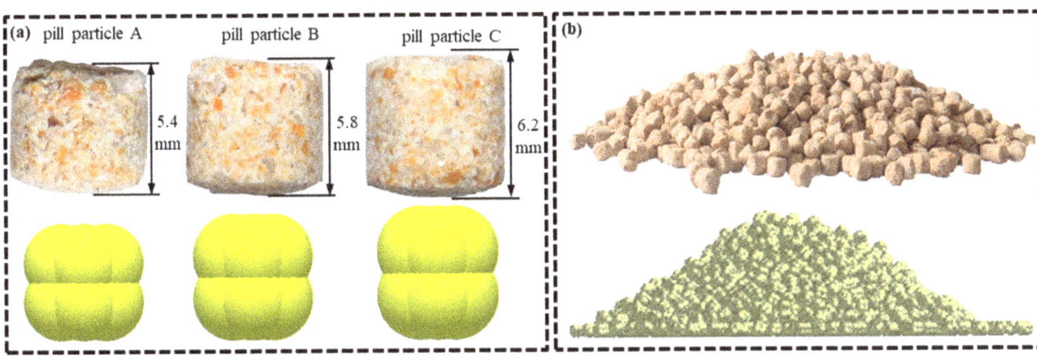

Figure 2. Models of pill particles and angle of repose test: (**a**) hree kinds of pill particles and the DEM models; (**b**) the angle of the repose test and simulation.

Based on an examination of the physical characteristics of the pill particles, parameters such as density, modulus of elasticity, and coefficient of static friction were calculated [24–27]. The coefficient of rolling friction was calibrated using the angle of repose tests, as shown in Figure 2b.

2.3. Simulation Contact Model and Parameters

To simulate the pill particle discharge process simulation test, the EDEM software Hertz–Mindlin contact mechanics model was employed [28–31]. According to the following model, the normal contact forces between the particles were estimated [20].

$$\mathbf{F}_{ij}^{n} = \left(\frac{4}{3}E^{*}\sqrt{R^{*}}\delta_{n}^{\frac{3}{2}} - 2\sqrt{\frac{5}{6}}\beta\sqrt{S_{n}m^{*}}(\mathbf{v}_{ij}\cdot\hat{\mathbf{n}}_{ij})\right)\hat{\mathbf{n}}_{ij} \tag{1}$$

where the equivalent Young's Modulus E^* and the equivalent radius R^* are defined as $E^* = \left[(1-v_i^2)/E_i + (1-v_j^2)/E_j\right]^{-1}$ and $R^* = [1/R_i + 1/R_j]^{-1}$, with E_i, v_i, and R_i and E_j, v_j, and R_j being the Young's Modulus, Poisson ratio, and Radius of particles i and j, respectively; δ_n is the normal overlap; the damping factor β, the normal stiffness S_n, and the equivalent mass m^* are given by $\beta = -\ln e/\sqrt{\ln^2 e + \pi^2}$, $S_n = 2E^*\sqrt{R^*\delta_n}$, and $m^* = [1/m_i + 1/m_j]^{-1}$, with e, m_i, and m_j being the coefficient of restitution and the mass of each particle in contact; $\mathbf{v}_{ij}$ is the relative velocity between particle i and j; the unit vector $\hat{\mathbf{n}}_{ij}$ is calculated as $\hat{\mathbf{n}}_{ij} = (\mathbf{R}_i - \mathbf{R}_j)/|\mathbf{R}_i - \mathbf{R}_j|$.

The Coulomb Moore friction theory was used to compute the tangential contact forces between the particles, which were derived using the following model.

$$\mathbf{F}_{ij}^{s} = \min\left(-S_t\delta_t - 2\sqrt{\frac{5}{6}}\beta\sqrt{S_t m^{*}}(\mathbf{v}_{ij}\cdot\hat{\mathbf{s}}_{ij}), \mu_s\mathbf{F}_{ij}^{n}\right)\hat{\mathbf{s}}_{ij} \tag{2}$$

where the tangential stiffness S_t is given by $S_t = 8G^*\sqrt{R^*\delta_n}$, with G^* being the equivalent shear modulus; δ_t is the tangential overlap; μ_s is the coefficient of static friction; $\hat{\mathbf{s}}_{ij}$ is the unit tangent vector.

Newton's second law was used to calculate particle translation:

$$m_i\frac{d\mathbf{v}_i}{dt} = \sum_j(\mathbf{F}_{ij}^{n} + \mathbf{F}_{ij}^{s}) + m_i\mathbf{g} \tag{3}$$

where $\mathbf{v}_i$ is the translational velocity of particle i; $\mathbf{g}$ is the gravitational acceleration.

Euler's equation was used to compute particle rotation:

$$I_i\frac{d\boldsymbol{\omega}_i}{dt} = \sum_j(\mathbf{R}_i \times \mathbf{F}_{ij}^{s} - \mu_r R_i|\mathbf{F}_{ij}^{n}|\hat{\boldsymbol{\omega}}_i) \tag{4}$$

where $\boldsymbol{\omega}_i$ and I_i are the angular velocity and moment of inertia of particle i, respectively. $\mathbf{R}_i$ is a vector running from the center of the particle to the contact point, with its magnitude equal to particle radius R_i. μ_r is the coefficient of rolling friction.

The parameters selected for the simulation of the pill discharge process are shown in Table 1. The elastic modulus was appropriately constricted and adjusted to 79 GPa to reduce simulation time consumption without impacting calculation accuracy. Other parameters were calibrated on this basis [4].

Table 1. Parameter selection for simulation.

Parameters	Pill Particle	ABS Plastic	Galvanized Steel
Density ρ, kg/m^3	1380	1050	7865
Poisson's ratio v	0.350	0.394	0.300
Elastic modulus E, Pa	1.100×10^8	3.189×10^9	7.900×10^{10}
Restitution coefficient e	0.201	0.299	0.305
Coefficient of static friction μ_s	0.466	0.577	0.511
Coefficient of rolling friction μ_r	0.080	0.120	0.070

2.4. Test and Simulation Methods

This research was divided into two sections: simulation model and parameter accuracy calibration, and discharge process simulation. The simulation model and parameter accuracy calibration test involved both a bench test and a simulation analysis of the discharge process. The simulation model and parameters were calibrated and validated by comparing the experiment data to the simulation results for various rotation speeds of the discharge wheel with the cumulative value of pill discharge.

With the grid number, grid width, and rotation speed of the discharge wheel as test factors and the pill particle filling amount, filling density, discharge variation coefficient, and compressive force as evaluation indicators, simulation tests of the discharge process were conducted using the calibrated and validated simulation model. Table 2 shows the components and levels of the simulation tests, which each included 40 sets and three repetitions. Finally, the mechanical behavior of the pill particles was observed and calculated, as well as the results of the discharge simulation experiments. The influence of the groove wheel's structural and operational properties on discharge performance was also disclosed, as was the ideal parameter combination.

Table 2. Components and levels of the simulation tests.

	Grid Number	Grid Width, mm	Rotation Speed, rpm
1	6	75	10
2	8	100	15
3	10	125	20
4	12	150	25
5			30

The following was the formula for calculating the discharge variation coefficient:

$$CV = \frac{\sqrt{\sum_{1}^{n}(M_i - \overline{M})/n}}{\overline{M}} \tag{5}$$

where M_i is the total mass of pill particles discharged per 0.5 s, and $\overline{M}$ is the average mass of each sample.

Using EDEM 2018 software, the Hertz–Mindlin (no-slip) contact model and the parameters provided in Table 1 were used to simulate the pill particle discharge processes. In proportions of 35%, 35%, and 30%, respectively, the particle models of the 12-sphere pills A, B, and C were used. The particle factory was first set up as a box area with dimensions of 400-mm-long by 150-mm-wide, with 5 mm above the top of the work bin. Second, the particle factory created 10 kg of almost 60,000 pill particles, which were then progressively accumulated in the hopper. The pill wheel began to spin as soon as the pill particles stabilized. Finally, a calculation was made to determine the assessment indications for the discharge process within 10 s.

3. Results and Discussion

3.1. Simulation Model and Parameter Validation

The width was 100 mm, and the number of grids was eight. The operating procedure of the discharge wheel was studied using an experiment and a simulation at 10 and 15 rpm.

The simulation and testing findings for pill discharge at various rotation speeds increased approximately linearly with time, as shown in Figure 3. When the speed of the discharge wheel was increased from 10 to 15 rpm, the linear deviation of the discharge test results decreased, showing that increasing the speed at lower rates improved the uniformity and stability of the pill-dispensing process. The simulation model and settings utilized in this research could accurately simulate the discharge process of pill particles, as shown by the comparison of simulation and experiment findings.

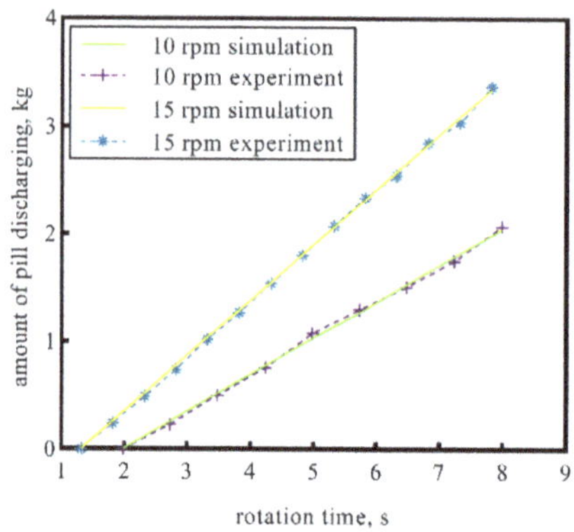

Figure 3. Experiment data and simulation results of the parameter validation.

3.2. Simulation Analysis of the Discharge Process

Figure 4 depicts a simulation examination of the pill discharge wheel's functioning procedure. The analysis of particle velocity is shown in Figure 4a. During the rotation of the discharge wheel, the pill particles could be split into three zones, with area I being the quasi-static zone, where the pill particles remained moderately static, gradually falling as the filling process continued. Area II was the driving zone, which was influenced by the dispensing wheel's rotation and had a particular relative velocity with the groove wheel, which had a significant impact on the agent's filling process. Area III was the filling zone, where particles collected in the grid due to gravity and were impacted by the grid's shape and size.

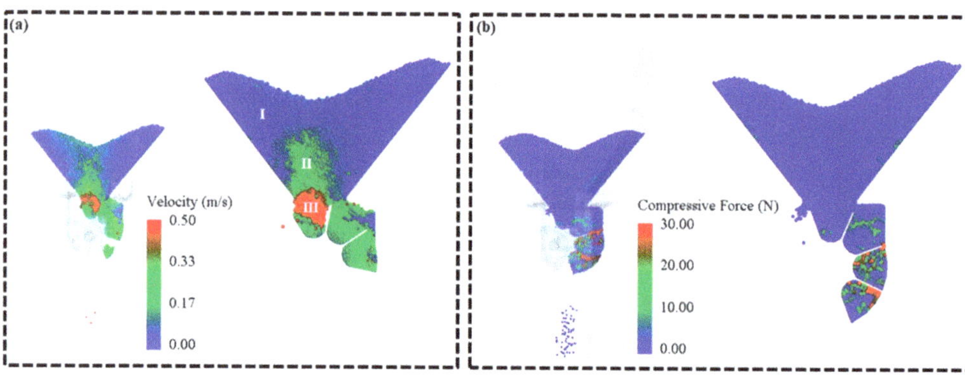

Figure 4. Particle velocity and compressive force analysis during pill filling and discharge process simulation: (**a**) the particle velocity analysis; (**b**) the particle compressive force analysis.

The examination of the particles' compressive force is shown in Figure 4b. The particles were prone to tight contact and jamming between the wheel and the outside wall during the rotation of the discharge wheel, resulting in a significant compressive force and particle breakage. The results reveal that the structure and operating factors, such as grid design and size, as well as the speed of wheel rotation, had a significant impact on the particle loading and discharge process and performance.

3.3. Effect of the Grid Number

The working process of the discharge wheel was researched for four grid numbers and five rotation speeds at a grid width of 100 mm in order to study the effect of the grid number on the discharge process and performance. Figure 5 depicts the simulation findings.

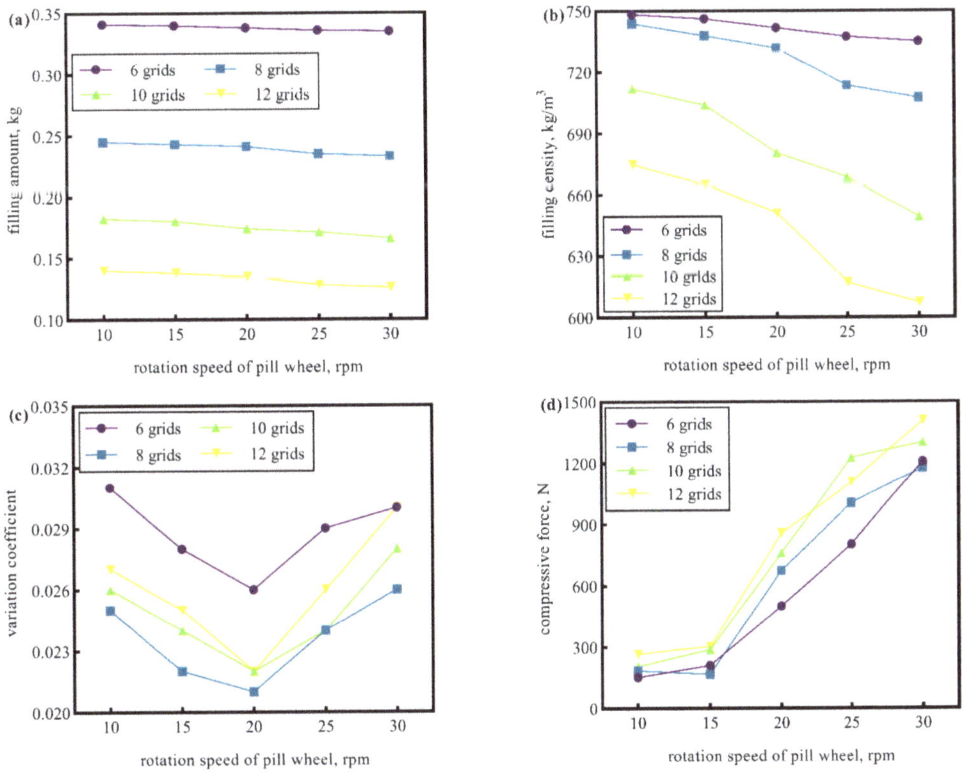

Figure 5. The effects of the grid numbers on the working performance over the rotation speed of the discharge wheel: (**a**) the filling amount; (**b**) the filling destiny; (**c**) the discharge variation coefficient; (**d**) the compressive force of the pill particles.

3.3.1. Effect on Filling Process

For varied grid numbers, Figure 5a demonstrates how the simulation results of the pill filling amount varied with the rotation speed of the discharge wheel. As the grid number grew, the average filling amount dropped from 0.34 to 0.13 kg, and the filling amount reduced as the rotation speed increased for different grid numbers. The filling capacity decreased as the grid numbers increased, but the total difference was minor.

Figure 5b depicts the relationship between filling density and discharge wheel rotation speed for various grid numbers. The change in filling density and the filling amount of the pill were essentially the same, as shown in the figure. For various grid numbers, the filling density fell as the rotation speed increased. When the number of grids was increased from 10 to 30 and the rotation speed was increased from 10 to 30 rpm, the filling density was reduced the most, from 674.78 to 607.31 kg/m^3.

3.3.2. Effect on Discharge Process

For varied grid numbers, Figure 5c demonstrates how the discharge variation coefficient varied with the rotation speed of the discharge wheel. When the number of grids was increased from 6 to 8, the variation coefficient lowered from 0.029 to 0.024, and when the number of grids was extended from 8 to 12, the variation coefficient decreased from 0.024 to 0.026. When a result, as the grid numbers increased, the stability of pill ejection increased and then declined.

3.3.3. Effect on Compressive Force

Figure 5d demonstrates how the compressive force of the particles varied with the rotation speed of the discharge wheel for various grid numbers. The average value of the compressive force increased from 574.20 to 787.24 N when the grid number was increased from 6 to 12. At various grid numbers, the compressive force of the particles increased dramatically as the rotation speed increased. When the number of grids was 12 and the discharge wheel rotation speed was 30 rpm, the maximum particle compressive force was 1407.50 N.

3.3.4. Discussion of Grid Number

The grid number in the discharge wheel was directly related to the shape and volume of the grid, according to the filling process study. As the grid number increased, the volume of the grids decreased, making it easier for the particles to form an arch, lowering the filling amount and density. The relative speed between the particles and the wheel increased as the rotation speed increased, affecting the filling process. As a result, the filling process and performance were affected by both the grid number and the rotation speed of the discharge wheel.

A critical factor for evaluating the discharge process was the consistency and consistency of particle discharge. When looking at the discharge findings, it can be seen that the rotation speed had a greater impact than the grid number. The pill particles could be ejected constantly and steadily once the rotation speed reached a specified range.

The impact of the discharge wheel's speed on the compressive force of the pill particles was stronger, according to the investigation. When the speed exceeded 15 rpm, the particles tended to become trapped between the trough wheel and the outer wall, resulting in a greater compressive force. Particle entrapment was more likely when the number of grids was large.

The results of the preceding investigation reveal that when the number of grids was 8, the discharge wheel filling and discharge process was improved, and the agent particles were less likely to break down at lower speeds.

3.4. *Effect of the Grid Width*

Based on the findings of the previous study, the working process and performance of the discharge wheel with a grid number of 8 were investigated for four grid widths and five rotation speeds, as shown in Figure 6.

3.4.1. Effect on Filling Process

Figure 6a demonstrates how the filling amount varied with rotation speed for various grid widths. As can be seen in the graph, the filling amount increased when the grid width grew; for example, as the grid width went from 75 to 150 mm, the average filling amount increased from 0.16 to 0.36 kg. For various grid widths, the filling amount dropped gradually as the rotation speed increased.

Figure 6b depicts the change in the pill filling density simulation results as a function of rotation speed for various grid widths. The average filling density increased in the order of 75, 150, 100, and 125 mm as the grid number increased in the figure, and the average filling density progressively increased from 666.35 to 726.88 kg/m^3. For varied grid widths, the filling density fell as the rotation speed increased. When the rotation speed was increased from 10 to 30 rpm at a trough wheel width of 75 mm, the filling density decreased the most, from 708.46 to 619.39 kg/m^3.

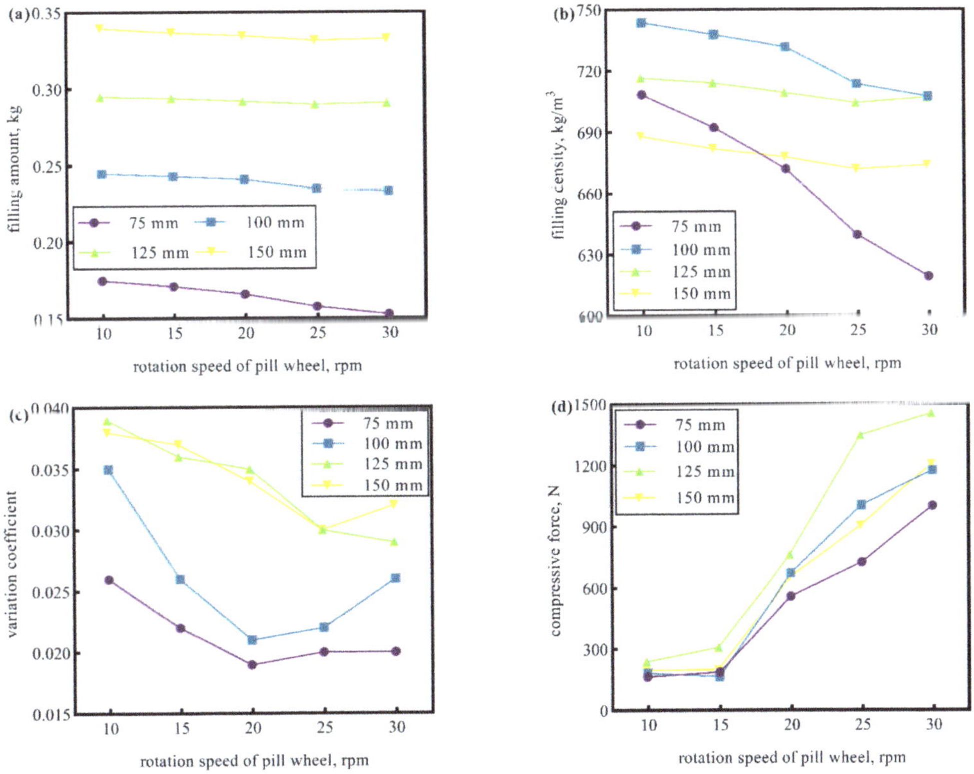

Figure 6. The effects of the grid widths on the working performance over the rotation speed of the discharge wheel: (**a**) The filling amount; (**b**) The filling destiny; (**c**) The discharge variation coefficient; (**d**) The compressive force of the pill particles.

3.4.2. Effect on Discharge Process

Figure 6c depicts the variation coefficient as a function of discharge wheel rotation speed for various grid widths. The variation coefficient rose from 0.0214 to 0.0342 as the grid width increased. For grid widths of 75 and 100 mm, the variation coefficient dropped and subsequently climbed, reaching its lowest value at 15 rpm. For grid widths of 125 and 150 mm, the variation coefficient was lower, and it alternated with rotation speed, reaching a minimum at 30 rpm.

3.4.3. Effect on Compressive Force

For varying grid widths, Figure 6d shows how the compressive force of pill particles varied with the rotation speed of the discharge wheel. With the increase in grid width and rotation speed, the average compressive force of the particles increased steadily from 528.2 to 634.4 N. At a grid width of 125 mm and a rotation speed of 30 rpm, the maximum particle compressive force was 1456.24 N.

3.4.4. Discussion of Grid Number

When the number of grids was 8, the filling density increases with the grid width, but this effect gradually diminished as the grid width went to a particular range, according to the filling process study. The filling density reduced as the wheel rotated faster, and this impact grew as the wheel rotated faster.

The variation coefficient was low, and the pill discharge was more uniform when the grid number was 8 and the grid widths were 75 and 100 mm, according to the results of the discharge process. The pill discharge process performed best when the rotation speed was between 10 and 15 rpm, which corresponds to the existing structural specifications of the discharge wheel. As a result, the grid width had a bigger influence on the total pill dispensing procedure.

The compressive force of the pill particles was calculated to illustrate that as the grid width rose, the possibility of the pill particles becoming trapped between the wheel and the outer wall increased, resulting in a higher compressive force. As a result, increasing the grid width increased the risk of pill particle fracture.

The performance of the filling amount, filling density, discharge variation coefficient, and compressive force of the particles during the operation of the discharge wheel was optimal when the grid number was 8, the grid width was 75 mm, and the rotation speed of the discharge wheel was 15 rpm, according to a comprehensive analysis of the effects of the grid number and width.

4. Conclusions

This research presented a simulation analysis of the working performance of the pill discharge device and the mechanical behavior of the pill particles during the pill discharge process using DEM in order to optimize the process of casing annular corrosion in oil and gas fields. The Hertz–Mindlin contact mechanics model was used to study the working process of the discharge device, analyze the contact between pill particles and the discharge device, and analyze the flow process and mechanical behavior of pill particles, due to the non-adhesive characteristics of pill particles. We discovered that the model accurately reproduced the dynamic properties of the pill particle population during discharge device operation. The simulation model and its parameters could more correctly mimic and anticipate the operating process and performance of the discharge device when compared to the actual test results.

In this paper, it was discovered that increasing the number of grids and decreasing the grid width of the discharge wheel, while having a minor effect on the filling amount and filling density, could reduce the time of a single discharge in the discharge process, lowering the variation coefficient of the pill discharge and ensuring that the pill was discharged uniformly and steadily. The discharge variation coefficient tended to drop and then increase as the rotation speed of the discharge wheel increased, and the compressive force of the pill particles gradually increased. As a result, lowering the rotation speed of the discharge wheel could increase pill discharge stability while also reducing particle breakage. The study's findings revealed that the best combination of grid number, grid width, and discharge wheel rotation speed was 8, 75 mm, and 15 rpm.

The findings of this research can be used to develop novel simulation models, design theories, and parametric guidance for future wellbore casing annular hollow corrosion protection agent addition devices in oil and gas fields, considerably simplifying the design process. It offers the groundwork for quickening the process of changing the way oil and gas field wellbore casing annular hollow corrosion protection is handled, as well as increasing oil and gas field wellbore integrity management capabilities, among other things.

Author Contributions: Conceptualization, D.L. and Y.Y.; methodology, D.L. and Y.Y.; software, Y.Y.; validation, Y.L. and H.L.; formal analysis, C.Q.; investigation, Z.Y.; resources, Y.L.; data curation, K.D.; writing—original draft preparation, J.S.; writing—review and editing, D.L.; visualization, S.C.; supervision, H.L.; project administration, Y.L.; funding acquisition, D.L. All authors have read and agreed to the published version of the manuscript.

Funding: The research was financially supported by Scientific Research Project of the CNOOC EnerTech-Drilling & Production Co., grant number GCJSXMHT-T2101.

Institutional Review Board Statement: Not applicable.

Informed Consent Statement: Not applicable.

Data Availability Statement: Not applicable.

Conflicts of Interest: The authors declare no conflict of interest.

References

1. Qiu, Z.C.; Xiong, C.M.; Ye, Z.R.; Yi, R.; Zhang, N. Wellbore anti-corrosion technique research in B block on the right bank of Amu Darya river sour gas field. *Anti-Corros. Methods Mater.* **2019**, *66*, 67–73. [CrossRef]
2. Laumb, J.D.; Glazewski, K.A.; Hamling, J.A.; Azenkeng, A.; Watson, T.L. Wellbore corrosion and failure assessment for CO_2 EOR and storage: Two case studies in the Weyburn field. *Int. J. Greenh. Gas Control* **2016**, *54*, 479–489. [CrossRef]
3. Chasemi, M.F.; Mehrpooya, M.; Chiasi, M.M.; Mohammadi, A.H.; Zendehboudi, S. Regional tectonic state and poro-thermo-elasticity analysis of near wellbore zone in field development plan: Utilization of an uncoupled approach. *J. Nat. Gas Sci. Eng.* **2017**, *46*, 615–636.
4. Liu, D.; Qiao, C.; Wan, J.; Lu, Y.; Song, J.; Yao, Z.; Wei, X.; Yu, Y. Modelling method and application of anti-corrosion pill particles in oil and gas field wellbore casing annulus based on the discrete element method. *Processes* **2022**, *10*, 1164. [CrossRef]
5. Zhou, H.; Chen, Y.; Li, H.; Xu, Z.; Dong, H.; Wang, W. Effect of particles micro characteristics destroyed by ball milling on fly ash electrostatic separation. *Adv. Powder Technol.* **2022**, *33*, 103449. [CrossRef]
6. Cundall, P.A.; Strack, O.D.L. Discrete numerical model for granular assemblies. *Géotechnique* **1979**, *29*, 47–65. [CrossRef]
7. Stoimenov, N.; Ruzic, J. Analysis of the particle motion during mechanical alloying using EDEM software. *IFAC-PapersOnLine* **2019**, *52*, 462–466. [CrossRef]
8. Langston, P.A.; Tuzun, U.; Heyes, D.M. Continuous potential discrete particle simulations of stress and velocity-fields in hoppers—Transition from fluid to granular flow. *Chem. Eng. Sci.* **1994**, *49*, 1259–1275. [CrossRef]
9. Yang, L.W.; Chen, L.S.; Zhang, J.Y.; Liu, H.J.; Sun, Z.C.; Sun, H.; Li, Y.D.; Zheng, L.H. Fertilizer sowing simulation of a vari-able-rate fertilizer applicator based on EDEM. *IFAC-PapersOnLine* **2018**, *51*, 418–423. [CrossRef]
10. Yan, Y.F.; Meng, D.X.; Song, Z.H.; Liu, L.Q.; Li, F.D. Particle kinetic simulation and experiment for flute-wheel feeding machine. *Trans. Chin. Soc. Agric. Mach.* **2016**, *47*, 249–253.
11. Han, D.; Zhang, D.; Jing, H.; Yang, L.; Cui, T.; Ding, Y.; Wang, Z.; Wang, Y.; Zhang, T. DEM-CFD coupling simulation and optimization of an inside-filling air-blowing maize precision seed-metering device. *Comput. Electron. Agric.* **2018**, *150*, 426–438. [CrossRef]
12. Ni, S.-H.; Yang, Y.-Z.; Huang, Y.-H. An EMD-based procedure to evaluate the experimental dispersion curve of the SASW method. *J. Chin. Inititute Eng.* **2014**, *37*, 883–891. [CrossRef]
13. Peschaid, I. Forging model/world relations: Relevance and reliability. *Philos. Sci.* **2012**, *79*, 749–760. [CrossRef]
14. Chen, E.J.; Kelton, W.D. Empirical evaluation of data-based density estimation. In Proceedings of the 2006 Winther Simulation Conference, Monterey, CA, USA, 3–6 December 2006; pp. 332–340.
15. Zhang, X.H. Computer Numerical Simulation of Pellet Feed in Cooling Process. Master's Thesis, Huazhong Agricultural University, Wuhan, China, 2004.
16. Yu, Y.; Fu, H.; Yu, J. DEM-based simulation of the corn threshing process. *Adv. Powder Technol.* **2015**, *26*, 1400–1409. [CrossRef]
17. Tekeste, M.Z.; Way, T.R.; Syed, Z.; Schafer, R.L. Modeling soil-bulldozer blade interaction using the discrete element method (DEM). *J. Terramechanics* **2020**, *88*, 41–52. [CrossRef]
18. Zhou, L.; Yu, J.; Wang, Y.; Yan, D.; Yu, Y. A study on the modelling method of maize-seed particles based on the discrete element method. *Powder Technol.* **2020**, *374*, 353–376. [CrossRef]
19. Langston, P.; Tüzün, U.; Heyes, D. Discrete element simulation of granular flow in 2D and 3D hoppers—dependence of discharge rate and wall stress on particle interactions. *Chem. Eng. Sci.* **1995**, *50*, 967–987. [CrossRef]
20. Langston, P.; Tüzün, U.; Heyes, D. Discrete element simulation of internal-stress and flow-fields in funnel flow hoppers. *Powder Technol.* **1995**, *85*, 153–169. [CrossRef]
21. Zhu, H.P.; Zhou, Z.Y.; Yang, R.Y.; Yu, A.B. Discrete particle simulation of particulate systems: Theoretical developments. *Chem. Eng. Sci.* **2007**, *62*, 3378–3396. [CrossRef]
22. Zhou, Y.C.; Wright, B.D.; Yang, R.Y.; Xu, B.H.; Yu, A.B. Rolling friction in the dynamic simulation of sandpile formation. *Phys. A Stat. Mech. Appl.* **1999**, *269*, 536–553. [CrossRef]
23. Zhu, H.P.; Yu, A.B. Averaging method of granular materials. *Phys. Rev. E* **2002**, *66*, 021302. [CrossRef]
24. Yan, D.; Yu, J.; Wang, Y.; Zhou, L.; Yu, Y. A general modeling method for soybean seeds based on the discrete element method. *Powder Technol.* **2020**, *372*, 212–226. [CrossRef]
25. Xu, T.; Yu, J.; Yu, Y.; Wang, Y. A modeling and verification approach for soybean seed particles using the discrete element method. *Adv. Powder Technol.* **2019**, *29*, 3274–3290. [CrossRef]
26. Zhou, L.; Yu, J.; Liang, L.; Yu, Y.; Yan, D.; Sun, K.; Wang, Y. Study on key issues in the modeling of maize seeds based on the multi-sphere method. *Powder Technol.* **2021**, *394*, 791–812. [CrossRef]
27. Horabik, J.; Wiącek, J.; Parafiniuk, P.; Bańda, M.; Kobyłka, R.; Stasiak, M.; Molenda, M. Calibration of discrete-element-method model parameters of bulk wheat for storage. *Biosyst. Eng.* **2020**, *200*, 298–314. [CrossRef]
28. Tavares, L.M.; Rodriguez, V.A.; Sousani, M.; Padros, C.B.; Ooi, J.Y. An effective sphere-based model for breakage simulation in DEM. *Powder Technol.* **2021**, *392*, 473–488. [CrossRef]

29. Neto, A.G.; Hudobivnik, B.; Moherdaui, T.F.; Wriggers, P. Flexible polyhedra modeled by the virtual element method in a discrete element context. *Comput. Methods Appl. Mech. Eng.* **2021**, *387*, 114163. [CrossRef]
30. Zeng, Y.; Mao, B.; Jia, F.; Han, Y.; Li, G. Modelling of grain breakage of in a vertical rice mill based on DEM simulation combining particle replacement model. *Biosyst. Eng.* **2022**, *215*, 32–48. [CrossRef]
31. Tao, H.; Zhong, W.; Jin, B. Comparison of construction method for DEM simulation of ellipsoidal particles. *Chin. J. Chem. Eng.* **2013**, *21*, 800–807. [CrossRef]

processes MDPI

Article

Process Development for Methyl Isobutyl Ketone Production Using the Low-Pressure One-Step Gas-Phase Selective Hydrogenation of Acetone

Abdulrahman A. Al-Rabiah [1,*], Raed R. Alkathiri [1] and Abdulaziz A. Bagabas [2]

1 Chemical Engineering Department, King Saud University, P.O. Box 800, Riyadh 11421, Saudi Arabia
2 National Petrochemical Technology Center (NPTC), King Abdulaziz City for Science and Technology (KACST), P.O. Box 6086, Riyadh 11442, Saudi Arabia
* Correspondence: arabiah@ksu.edu.sa; Tel.: +966-11-4676844; Fax: +966-11-4678770

Abstract: Methyl isobutyl ketone (MIBK) is a highly valuable product in the chemical industry. It is widely used as an extracting agent for heavy metals, antibiotics, and lubricating oils. Generally, MIBK can be produced by three-step and one-step liquid-phase methods. These methods are expensive and energy-demanding due to the high pressure and low conversion of acetone. A novel nano-Pd/nano-$ZnCr_2O_4$ catalyst was developed to produce MIBK with high conversion and selectivity of 77.3% and 72.1%, respectively, at 350 °C and ambient pressure, eliminating the need for high pressure in conventional MIBK processes. This study is the first that proposes a newly developed process of methyl isobutyl ketone (MIBK) production using the low-pressure one-step gas-phase selective hydrogenation of acetone. In this work, a novel process flow diagram has been developed for the production of MIBK using the developed nano-catalyst. The process was heat integrated, resulting in a 26% and a 19.5% reduction in the heating and cooling utilities, respectively, leading to a 12.6% reduction in the total energy demand. An economic analysis was performed to determine the economic feasibility of the developed process, which shows that the process is highly profitable, in which it reduced both the capital and operating costs of MIBK synthesis and showed a return on investment (ROI) of 29.6% with a payback period of 2.2 years. It was found that the ROI could be increased by 18% when the reactor temperature is increased to 350 °C. In addition, the economic sensitivity analysis showed that the process is highly sensitive to product prices and least sensitive to utility prices, which is due to the versatility of the process that requires only a low amount of energy.

Keywords: methyl isobutyl ketone (MIBK); acetone self-condensation; selective hydrogenation; process development; heat integration; economic analysis; nano-Pd/nano-$ZnCr_2O_4$ catalyst

Citation: Al-Rabiah, A.A.; Alkathiri, R.R.; Bagabas, A.A. Process Development for Methyl Isobutyl Ketone Production Using the Low-Pressure One-Step Gas-Phase Selective Hydrogenation of Acetone. *Processes* **2022**, *10*, 1992. https://doi.org/10.3390/pr10101992

Academic Editor: Jean-Claude Assaf

Received: 1 September 2022
Accepted: 27 September 2022
Published: 2 October 2022

1. Introduction

Methyl isobutyl ketone (MIBK) is an industrially important chemical worldwide. It can be used in a wide variety of important applications, including protective surface coating, vinyl, and acrylic resins, the extraction of metallic salts, and the removal of paraffin from mineral oils [1–5]. It is also used in the production of paints and rubber, as well as a solvent in a variety of pharmaceuticals and adhesives [6,7].

The global production of MIBK was 430,000 metric tons in 2020 [8]. The market for MIBK is expected to grow at a compound annual growth rate (CAGR) of more than 6% by 2026. This is due to the increasing demand for MIBK, especially in the synthesis of antioxidant additives for rubber, and the growing demand for solvents in chemical processes [9].

In addition to MIBK, a crucial byproduct is also produced, which is isopropyl alcohol (IPA). It has tremendous industrial, medical, and hygienic applications, including alcohol wipes, hand sanitizers, disinfectants, and swabs, which increases the prospects of applying the MIBK process with minimal waste. According to the Centers for Disease Control and

Prevention (CDC) in the United States, alcohol-based hand sanitizers and cleansing wipes contain between 60% and 90% of alcohol, and the use of 70% of IPA is recommended [10]. Medical IPA was scientifically proven as a disinfectant that inactivates viruses such as the COVID-19 virus.

MIBK can be produced by several processes, but essentially by a three-step [11] method and a single-step method [12–17]. Each of these processes can be carried out with different catalysts, raw materials, and reaction conditions [12]. The three-step method involves aldol condensation of acetone over homogeneous aqueous-based catalysts, such as sodium and calcium hydroxide, forming diacetone alcohol (DAA), which is subsequently dehydrated to mesityl oxide (MO). The final step is the selective hydrogenation of the C=C bond of MO to MIBK using Cu or Ni catalysts [18,19]. The most commonly occurring side reactions are over-condensation and unselective hydrogenation [20].

The three-step process produces a considerable amount of wastewater containing acidic impurities due to the homogeneous catalyst, which creates a corrosive environment associated with high capital and operating costs for the neutralization process, acetone condensation, and product purification [18,19,21]. Alternatively, the one-step process using a metal-supported catalyst was demonstrated to be more efficient [22–28]. The support catalyzes acetone condensation to mesityl oxide (MO), an α,β-unsaturated ketone, followed by in situ catalytic reductions to MIBK using palladium on acidic ion-exchange resins [18,29,30], while there are other processes that use Pd/zirconium phosphate [31]. In the one-step liquid-phase process, hydrogen and acetone are passed over metal base catalysts at moderate temperature and high pressure to produce MIBK in a single step, saving enormous costs that are required in the three-step method [32]. Under these reaction conditions, aldol condensation and dehydration are reversible [33], but the catalyst shifts the equilibrium toward MO, irreversibly hydrogenating it toward MIBK [28].

Despite the many advantages offered by the liquid-phase one-step process, it suffers from high pressure, ranging from 3 to 10 MPa [13], which significantly increases the capital and operating costs. This is in addition to the large recycle flow rate due to the low conversion of acetone, which increases the size of equipment in the recycle loop as well as the additional energy that is needed for recovering the unreacted acetone in the distillation column. Therefore, it is an industrial objective to reduce the pressure and increase the conversion and selectivity to make the process more profitable.

Recently, Qianling et al. reported a bifunctional catalyst TiO_2/Al_2O_{32} and Pd/Cor for one-pot liquid-phase synthesis of MIBK from acetone, which provides acetone conversion of (35–45)% with a selectivity of (80–90)% to MIBK at 150 °C and 2.0 MPa [34]. However, the process still suffers from the high-pressure requirement with moderate acetone conversion. Baining et al. proposed an alternative support with high thermal stability and adjustable acid-base property using phosphor-doped h-BN that catalyzes the aldol condensation and dehydration to provide an acetone conversion of 58.24% and selectivity of 68.39% to MIBK [35]. In addition, various kinds of support, including metal oxides [19,36,37] and sulfonated graphene [38,39], have also been a prime focus in the development of the process.

In this context, a novel nanocrystalline zinc chromite-supported nanopalladium (nano-Pd/nano-$ZnCr_2O_4$) catalyst was developed by Bagabas et al. and Al-Rabiah et al. [20,40,41], which enables the production of MIBK in a one-step gas-phase reaction at atmospheric pressure. This can overcome the infirmities that the conventional process suffers from. However, the process requires a temperature of (200–350) °C, which is higher than the conventional process, which uses a temperature of (120–160) °C [15].

The reaction involves base–acid coupling of acetone to form mesityl oxide (MO), followed by its hydrogenation to methyl isobutyl ketone (MIBK). The products with this catalyst are methyl isobutyl ketone (MIBK) as the main product, isopropyl alcohol (IPA) as the main byproduct, and diiasobutyl ketone (DIBK), mesityl oxide (MO), and mesitylene (M) are also formed [20,40,41]. The MIBK produced by low-pressure one-step gas-phase acetone condensation has an acetone conversion of (20–78)% and MIBK selectivities of

(40–73)% for reaction temperatures of 200–350 °C with a hydrogen-acetone molar feed ratio of 1 to 6 [20,40,41].

This research work focuses on the development of a novel process for the production of methyl isobutyl ketone (MIBK) from hydrogen and acetone in a one-step gas-phase reaction at ambient pressure using a novel nano-Pd/nano-$ZnCr_2O_4$ catalyst. Heat integration is conducted to reduce the process utilities. In addition, this work investigates process economics to measure the profitability and the impact of market volatility on the profitability of the process.

2. MIBK Process Development

The developed process flow diagram for the production of methyl isobutyl ketone is shown in Figure 1. A process simulation was conducted using CHEMCAD v.7.1 software based on a production of 30,000 metric tons of MIBK per annum with 30 days assigned for maintenance. The thermodynamic model chosen was UNIFAC as a cubic equation of state. The reactor conversion, selectivity, and feed ratio used in the simulation were based on the aforementioned experiment.

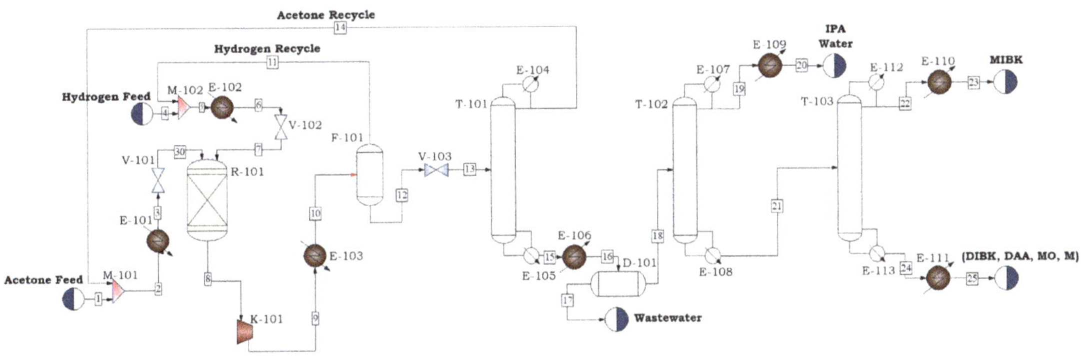

Figure 1. Developed process flow diagram (PFD) of the MIBK process.

The fresh acetone feed (stream 1) enters the process at 70 °C and 1.8 atm in a liquid phase and is mixed with the recycled acetone (stream 14), which enters at 74 °C and 1.8 atm. Before entering the reactor R-101, the mixed stream is heated to 300 °C. The hydrogen feed is mixed with recycled hydrogen (stream 11) before being heated to 300 °C and fed into reactor R-101. Both reactor feeds pass through valves, V-101 and V-102, to reduce the pressure in the reactor. The hydrogen–acetone molar feed ratio is maintained at 2:1, as recommended [40,41]. The reaction proceeds isothermally in the gas phase at 300 °C and 1 atm in a fixed bed catalytic reactor (R-101). The conversion of acetone is 66%, with a selectivity of 69.4% to methyl isobutyl ketone (MIBK). Other products, such as isopropanol (IPA) and diiasobutyl ketone (DIBK), are also produced. The selectivity of each of these substances was determined based on the experimental work, which is shown in Table 1. The reactor effluent will then be compressed to 6.5 atm to compensate for pressure losses through the pipelines and to enhance the separation of hydrogen without major product losses from the process stream at low pressure and high temperature, preventing the use of cryogenic conditions. MIBK losses at 1 atm are approximately 6 kmol/h. Consequently, compression is mandatory to avoid cryogenic conditions. At 6.5 atm, the losses drop to about 0.7 kmol/h of MIBK. After compression, the stream is cooled to 35 °C in E-103 before being fed into the flash drum F-101. The overhead stream from flash drum F-101 contains nearly 99.996% of the unreacted hydrogen, which is recycled and mixed with the fresh hydrogen. The bottom stream from flash drum F-101 (stream 12) is then throttled to 1.8 atm in V-103 before entering the distillation column, T-101. Nearly 99% of the unreacted acetone is recovered in the overhead stream of T-101, using 48 stages, before being mixed with the fresh acetone. The acetone/water system forms an azeotropic mixture at higher compositions; therefore, about 0.04 of the entered water into the column is recovered into the top stream to maintain the mixture in the column below the azeotropic point. The bottom stream from T-101 is cooled to 70 °C in E-106 before being sent to a water decanter, D-101, where 90% of the water is removed. The second outlet stream from the decanter enters the distillation column, T-102, which consists of 34 stages, where 99.28% of the isopropanol (IPA) is recovered. The remaining water is also separated with IPA in the distillate stream, and the mixture is then cooled to 40 °C in E-109. The purity of IPA in the top stream of T-102 is almost 68%, which is highly recommended to be used as sanitizer in the medical industry, and therefore, there is no need to install a further distillation column for increasing the purity of the IPA. The bottom stream from T-102 enters another distillation column, T-103, which consists of 67 stages where 99% of the MIBK is recovered in the overhead stream with a purity of 99.83 wt.%, achieving the product specifications [14]. The MIBK stream is then cooled to 40 °C in E-110 before being sent to the storage tank. The bottom stream from T-103, which contains the heavy components (e.g., DAA, MO, M, and DIBK), is cooled to 40 °C in E-111.

Table 1. The conversion and selectivity of the MIBK process at different temperatures [20,40,41].

Temperature (°C)	Acetone Conversion%	Selectivity					
		MIBK	DIBK	MO	M	IPA	DA & Others
200	20.1	40.6	10.2	6.1	2.1	40.7	0.3
250	40.7	53.9	12.3	4.4	2.5	25.8	1.1
300	66	69.4	12.4	2.1	2.6	11.2	2.3
350	77.3	72.1	13.5	2.2	2.7	5.4	4.1

3. Heat Integration in the MIBK Process

Heat integration is an essential part of the process development. The reduction in energy usage not only reduces the operating costs but also prevents harmful carbon emissions in the environment. The required heating and cooling utilities were identified for the process as shown in Figure 1. Each process stream was drawn with a tail and a head representing the supplied temperature and the target temperature, respectively. The final heat exchanger network is shown in Figure 2, where the red and blue lines denote the hot and cold streams, respectively. The constructed network utilizes the energy content and high temperature of the reactor effluent to heat up the feed streams. The acetone feed (stream 2) leaving mixer M-101 will first be heated by the reactor effluent from 74.3 °C to 242 °C, in which no further heating can occur due to the process constraint prohibiting phase change before the inlet of the compressor. The reactor effluent will be further cooled by the hydrogen feed (stream 5) until it reaches a minimum temperature approach of 10 °C to avoid an excessive large area. The outcome of the heat integration, as shown in Figure 3, is a reduction in the heating and cooling utilities by 26% and 19.5%, respectively. The reduction in the total utility demand is about 12.6%. In terms of economics, the implemented heat integration has resulted in an increase in the annualized capital cost by $12,134/year (for 10 years of operation). However, operating costs have decreased by $611,852/year, indicating a highly plausible energy utilization that satisfies both technical and economic aspects.

The blue columns in Figure 3 represent the utility demand before the heat integration, while the green columns represent the current value of the utility demand in the MIBK process after heat integration. In addition, the figure is divided into three sections, starting from the total utility demand, including the combined heating and cooling utilities, then the heating utility in the middle, while the cooling utility is placed on the right side. The process was re-simulated after the heat integration using CHEMCAD v.7.1. The developed flowsheet after heat integration is shown in Figure 4. The material balances for the key streams of the developed MIBK process are shown in Table 2.

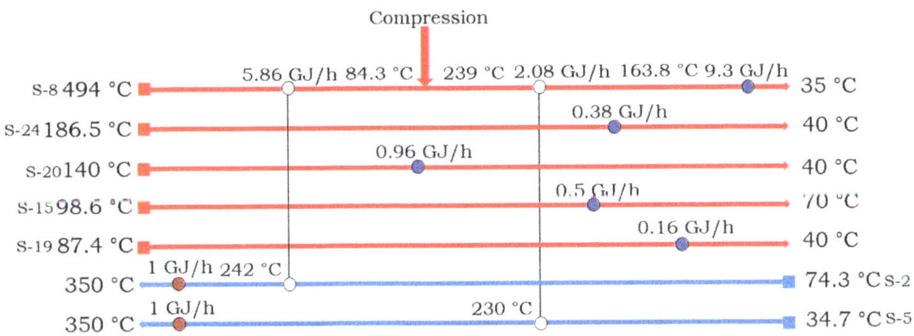

Figure 2. Heat exchanger network of the process streams for the developed process.

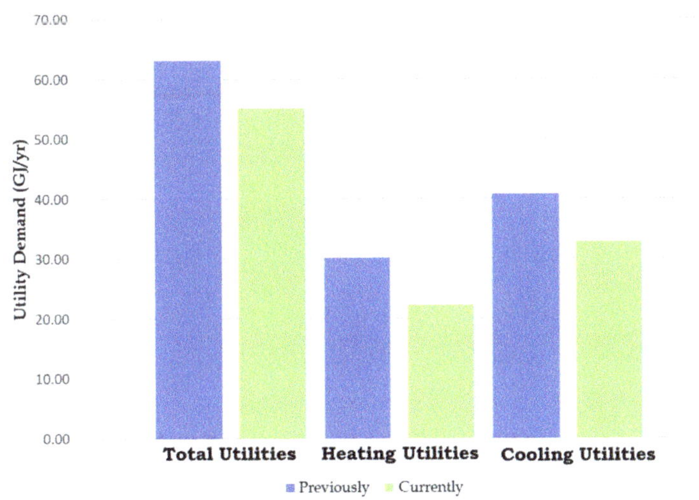

Figure 3. Heat integration outcome for the developed process.

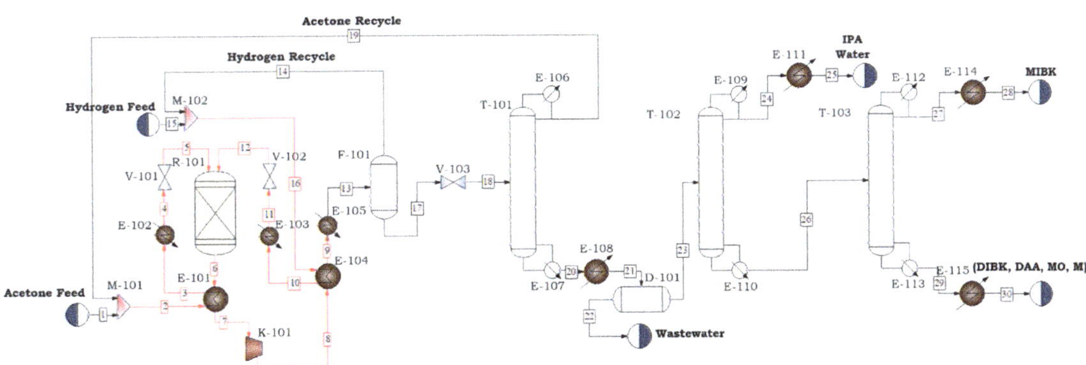

Figure 4. Developed process flow diagram (PFD) of the MIBK process after heat integration.

Table 2. Material balance of key streams for the developed MIBK process.

Streams	Acetone Mixed Feed (S-2)	Hydrogen Mixed Feed (S-16)	Reactor Effluent (S-6)	Acetone Recycle (S-19)	Hydrogen Recycle (S-14)	Waste-Water (S-22)	(IPA + Water) Product (S-25)	MIBK Product (S-28)	(DIBK, DAA, MO, M) Product (S-30)
Temperature (°C)	74.29	34.74	300	61.49	35	70	40	40	40
Pressure (atm)	1.8	5.9	1	1.8	5.9	1.4	1.4	1.9	2.4
Enthalpy (MJ/h)	−37,473.6	−2720.4	−35,506.3	−10,788	−2695.83	−12,968.1	−5515	−12,209.8	−3039.6
Molar vapor fraction	0.29	1	1	1	1	0	0	0	0
Molar flow rate (kmol/h)	160.11	342.63	443.12	50.13	283.34	45.94	18.02	37.46	8.25
Mass flow rate (kg/h)	9224.8	1266.14	10,491	2837.35	1146.63	827.54	−5515	3748.6	1054.04
				Component flow rates (kmol/h)					
Hydrogen	0.0109	331.31	272	0.0109	272	0	0	0	0
Acetone	158.3	7.64	56.4	48.27	7.64	0	0.49	0	0
MIBK	0	0.75	38.75	0	0.75	0	0.2492	37.37	0.38
Water	1.84	2.36	55.24	1.84	2.36	45.94	5.1	0	0
DAA	0	0.0012	1.26	0	0.0012	0	0	0	1.26
MO	0	0.0115	1.16	0	0.0115	0	0	0.0065	1.15
M	0	0.0148	0.9637	0	0.0148	0	0	0	0.9489
IPA	0.00645	0.5225	12.79	0.0065	0.5225	0	12.18	0.0883	0
DIBK	0	0.0155	4.54	0	0.0155	0	0	0	4.53

4. Economic Analysis

An economic analysis was performed to verify the profitability of the process based on several economic indicators, including the return on investment, payback period, net present worth, and discounted cash flow rate of return.

The return on investment (ROI) is expressed on an annual basis, and it is defined as the average net profit divided by the total fixed capital investment as shown in Equation (1) [42].

$$\text{ROI} = \frac{\text{Yearly net profits}}{\text{Fixed capital investement}} \tag{1}$$

The payback period (PBP) is defined as the required time to recoup the original depreciable fixed capital investment as shown in Equation (2) [42].

$$\text{PBP} = \frac{\text{Depreciable fixed capital investment}}{\text{Avg. profit/yr} + \text{avg. depreciation/yr}} \tag{2}$$

The net present worth (NPW) is defined as the difference between the present value of the annual cash flow and the initial required investment as shown in Equation (3).

$$\text{NPW} = \frac{\text{Net cash flow at time t}}{(1 + \text{interest rate})^{\text{time of cash flow}}} \tag{3}$$

The discounted cash flow rate of return (DCFRR) represents the maximum interest rate that a project can afford to pay for its total capital investment [43] as shown in Equation (4) [42].

$$\text{NPW} = \sum_{n=1}^{n=T} \frac{\text{Cash flow at time t}}{(1 + \text{DCFRR})^{n}} - \text{Total capital investment} = 0 \tag{4}$$

In order to determine the economic indicators, the following should be calculated: (i) the cost of the equipment; (ii) the costs of raw materials and utilities; (iii) the total capital cost; and (iv) the total production cost.

4.1. Purchase Cost

In order to determine the purchase costs, the process equipment sizes must first be calculated; the process simulator was used for this purpose. In addition, the costs were estimated using the CHEMCAD costing package, adjusting all the costs to 2022 using the Chemical Engineering Plant Cost Index (CEPCI) [44]. Figure 5 depicts the distribution of the process equipment costs. Note that the condensers and reboilers of the distillation columns are compiled with heat exchangers. The cost distribution of the process equipment indicates that the distillation columns account for the largest share of fixed capital costs of around 40% as the most expensive units of the process, which is due to the byproducts generated by the process.

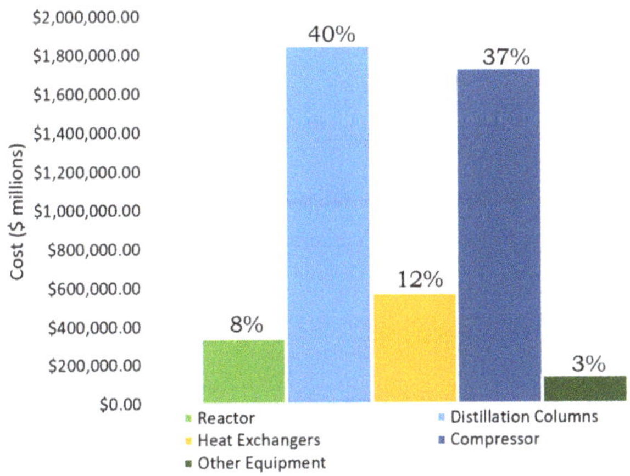

Figure 5. Fixed capital cost distribution for the developed process.

4.2. Raw Material and Utility Costs

Besides the fixed capital costs, the raw materials and utility costs are major components of the production cost. The prices of the raw materials, utilities, and products are shown in Table 3.

Table 3. Cost input for the developed MIBK process [45,46].

Item	Price Unit
Raw Materials	
Acetone	\$790/ton
Hydrogen	\$2500/ton
Utilities	
High-pressure steam (hps)	\$5.66/GJ
Medium-pressure steam (mps)	\$4.77/GJ
Low-pressure steam (lps)	\$4.54/GJ
Boiler feed water	\$1.532/1000 kg
Cooling water (cw)	\$15.7/1000 m^3
Electricity	\$0.0674/kWh
Products	
Methyl isobutyl ketone (MIBK)	\$2120/ton
Isopropanol (IPA)	\$1540/ton
(DAA, DIBK, MO, M)	\$1300/ton

The distribution of the utility costs is depicted in Figure 6. It shows that electricity represents 35% of the total utility cost, which is due to its high prices. Low-pressure steam is the second most cost-intensive required utility that is used in the reboiler of T-101. Although cooling water is the most consumed in terms of volume, it accounts for only 5%

of the total utility cost, which is due to the low price of cooling water compared to the other utilities.

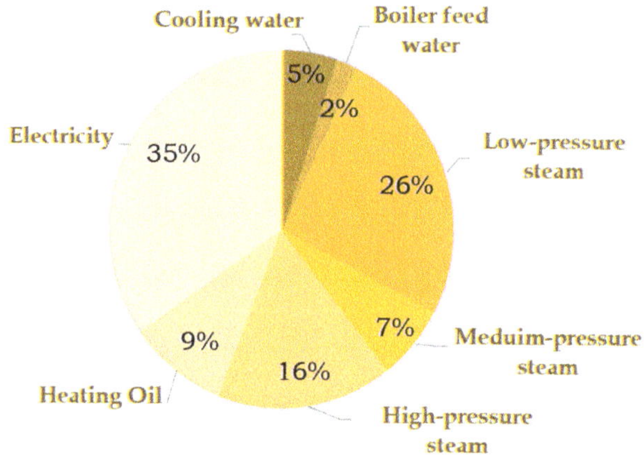

Figure 6. Utility cost distribution for the MIBK process.

4.3. Total Capital Cost

The total production cost comprises all of the direct, indirect, and general expenses [42]. The direct costs are estimated as a fraction of the purchased equipment costs and include all the prices of the delivery and installation of the purchased equipment, installed instrumentation and control systems, installed piping, electrical systems, yard improvement, buildings, and service facilities [42]. The direct cost of the developed MIBK process is $18.34 million per annum. The indirect cost was calculated as a fraction of the direct cost provided in Plant Design and Economics for Chemical Engineers by Peter and Timmerhaus [42]. It includes the engineering and supervision costs, construction, legal, contractor, and contingency expenses, which represent $7.34 million [42]. The fixed capital investment is the summation of the direct and indirect costs [42]; it is $25.68 million for the MIBK process. The total capital investment is a combination of the fixed capital investment and the working capital. The working capital was taken as 0.15 of the total capital cost for the fluid processing plant [42]. The total capital investment in the developed MIBK process is $30.21 million.

4.4. Total Product Cost

The total product cost represents all of the annual expenses that are required for production. It includes raw material prices, utility costs, operating labor, maintenance and repairs, operating supplies, laboratory charges, royalties, taxes, financing, insurance, rent, depreciation, plant overhead, administration, distribution and selling, and research and development expenses, all of which are fractions of the fixed capital investment provided by Peter and Timmerhaus [42]. The total production cost, excluding depreciation, for the developed MIBK process is $55.78 million annually.

4.5. Profitability Evaluation

The process profitability can now be evaluated after obtaining all of the necessary information. The total raw material cost is $42.71 million per year, which accounts for 76.7% of the total product cost. The total product price is $83.81 million yearly, which shows a positive profit margin. Therefore, fluctuations in the raw material or product prices may have a considerable influence on the process profitability; consequently, a sensitivity analysis is required for the price fluctuation. The plant has shown its feasibility in being commercially applied; the return on investment for the plant is 29.6%/y, with a payback period of 2.2 years. Furthermore, the plant could generate $28.93 million profits in 10 years of operation. The discounted rate of return on investment of the plant is 26.6%. Table 4 summarizes the economic performance of the MIBK process.

Table 4. The economic performance for the developed MIBK process.

Factor	Value
Total Raw Material Cost	$42.71 million/y
Total Utility Cost	$1.646 million/y
Fixed Capital Investment (FCI)	$25.68 million
Total Capital Investment (TCI)	$30.22 million
Depreciation	MACRS, 5 years recovery period
Total Product Cost	$55.78 million/y
Revenue	$83.81 million/y
Return on Investment (ROI)	29.6%/y
Payback Period	2.2 y
Net Present Worth (at i = 15%)	$28.93 million
Discounted Cash Flow Rate of Return (DCFRR)	26.6%

4.6. Economic Sensitivity Analysis

The petrochemical industry is a highly volatile market that has experienced many ups and downs throughout its market history [47,48]. Therefore, to account for these perturbations, raw material costs, product costs, utility costs, and fixed capital costs were varied by ±15% of their current prices. Figure 7 shows the overall sensitivity analysis of the above factors at a constant production rate and 10 years of operation. It shows that the process is highly dependent on the product prices, in which a 10% increase could make the process an extremely lucrative process with an ROI of 54.3%. However, a 15% decrease could make the process unprofitable with an ROI of 4.1%. The fluctuation in the product prices gave maximum and minimum values of ROI in the sensitivity analysis, which represents the importance of reducing the MIBK losses throughout the process. On the other hand, raw material prices come in as a second large contributor to the process profitability, in which a 15% increase in the raw material prices decreases the ROI to 12.7%, whereas a 15% reduction in the raw material prices increases the ROI to 46%. This highlights how crucial it is to properly recover unreacted raw materials (e.g., acetone and hydrogen), in addition to the great influence that the higher selectivity can have on the process profitability. The fixed capital investment was found to be the third largest contributor, with an ROI of 23.5% if its cost increased by 15% and an ROI of 37.7% if its cost decreased by 15%. This also highlights that even though recovering more of the product and raw materials can increase the TCI, it has a lower negative influence on the process profitability than losing them. In addition, it is worth noting that the utility cost, as expected, has an insignificant impact on the ROI as the current utility cost is $1.646 million per year, representing only 2.95% of the total product cost, which is due to the low-pressure requirement in the process compared to the conventional process, which is directly associated with the electricity consumption.

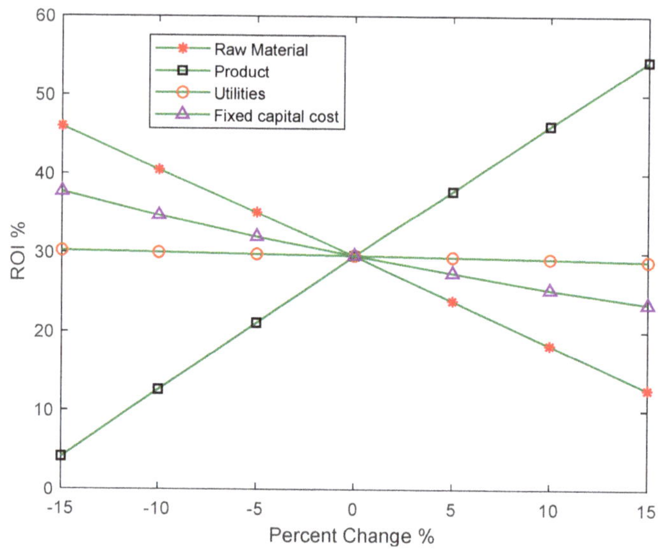

Figure 7. Return on investment under various market variations of the economic factors.

4.7. Effect of Reaction Temperature on the Process Profitability

The temperature for reactor operation proposed by Al-Rabiah et al. [40,41] is in the range of (200–350) °C. The conducted sensitivity analysis has shown that it is appealing to increase the conversion and selectivity of the process. In the current study, the used temperature is 300 °C. When the reactor temperature is increased to 350 °C, the conversion and selectivity are increased from 66% and 69.4% to 77.3% and 72.2%, respectively. Consequently, less acetone is required, and more MIBK is produced. Nevertheless, this also increases the heating requirement needed for the feed stream as well as the reactor cooling system. In order to determine the optimum temperature in terms of profitability, the process was simulated at 350 °C. All costs were determined in the same manner as previously done. Figure 8 shows that as the reactor temperature increases, the fixed capital investment decreases, which is due to the lower volume of the distillation column T-101 that is needed to separate acetone from the process stream. In addition, the rest of the columns' size remained almost the same as the production rate was fixed. In addition, the cost of the compressor decreased because the numbers of unreacted moles of acetone and hydrogen were lower. On the other hand, the heating utility that is required to maintain the reactor temperature was increased by about 10%. In addition, more heating utility was required for E-102 because the minimum temperature approach was increased from 10 °C to 30 °C to prevent a phase change prior to the compressor. It is worth noting that even though the inlet temperature to E-105 was increased from 164 °C to 180 °C, the heat duty was about 11% less due to the lower flow rate entering the heat exchanger. The condenser and reboiler heat loads of T-101 were also lower due to the lower amount of unreacted acetone. An overall assessment of energy consumption is shown in Figure 8, which indicates that increasing the temperature from 300 °C to 350 °C reduces the overall energy consumption of the process. Furthermore, the ROI increased to 34.3%, and the payback period decreased to 2 years.

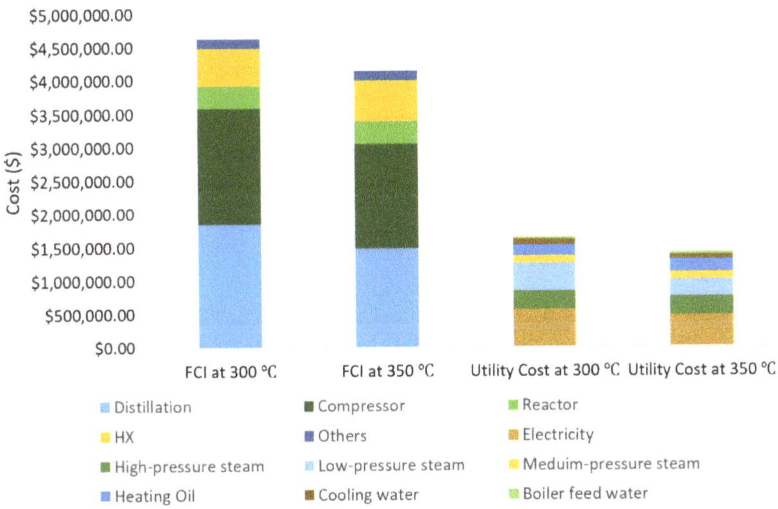

Figure 8. A comparison between the effect of the eactor temperature on the FCI and the utility cost.

In terms of profitability, the return on investment for the process when operating at 350 °C is 35.1%/y, which is 18% higher while the net present worth has upsurged by 12%. Figure 9 depicts the cash flow diagram of both process conditions.

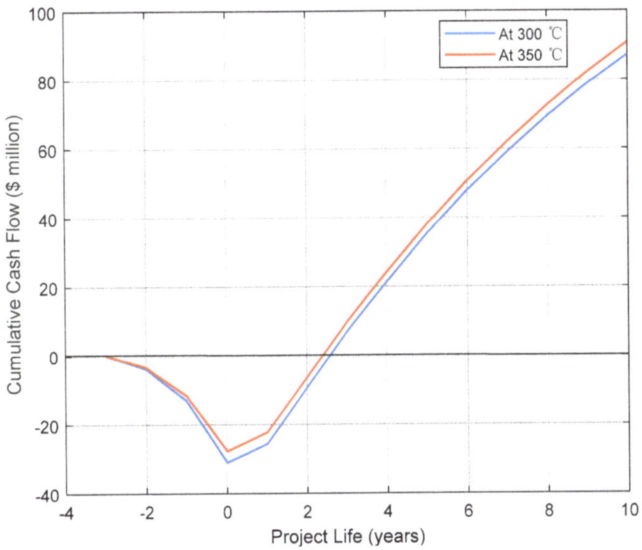

Figure 9. The cumulative cash flow diagram for both reactor temperatures.

5. Conclusions

Methyl isobutyl ketone (MIBK) is a chemical intermediate and an important solvent. It is used in several important applications in the chemical market. There are several methods to produce MIBK, either by the three-step process or by the one-step liquid-phase reaction method, which is commercially preferred. However, these methods require very high pressure, which is expensive and energy consuming. In contrast, a promising nano-Pd/nano-$ZnCr_2O_4$ catalyst was developed to produce MIBK in a one-step gas-phase reaction under ambient pressure, overcoming the drawback of the conventional process. In this work, a novel process flowsheet has been developed and simulated. The developed process was heat integrated by matching the process streams to deliver the surplus heat to the heat deficit streams, resulting in a 12.6% reduction in the total energy demand. A comprehensive economic analysis of the developed process was conducted and has shown that the process is highly profitable with a return on investment (ROI) of 29.6% and a payback period of 2.2 years. It is shown that when the process is operated at a temperature of 350 °C, the economics improved, and the ROI increased to 34.3%.

6. Patents

US Patent No. 7,951,976, 31 May 2011, US Patent No. 8,101,805, 24 January 2012, EP Patent No. 2,532,642 B1, 25 January 2017, EP Patent No. 2418017B1, 21 November 2018.

Author Contributions: Conceptualization, A.A.A.-R.; methodology, A.A.A.-R.; software, A.A.A.-R. and R.R.A.; validation, A.A.A.-R. and R.R.A.; formal analysis, A.A.A.-R. and R.R.A.; investigation, A.A.A.-R. and R.R.A.; resources, A.A.A.-R. and A.A.B.; data curation, A.A.A.-R. and A.A.B.; writing—original draft preparation, A.A.A.-R.; writing—review and editing, A.A.A.-R. and R.R.A.; visualization, A.A.A.-R. and R.R.A.; supervision, A.A.A.-R.; project administration, A.A.A.-R. and A.A.B.; funding acquisition, A.A.A.-R. and A.A.B. All authors have read and agreed to the published version of the manuscript.

Funding: This research received no external funding.

Data Availability Statement: Not applicable.

Acknowledgments: This project was supported by King Saud University, Deanship of Scientific Research, College of Engineering Research Center.

Conflicts of Interest: The authors declare no conflict of interest.

Abbreviations

Abbreviation	Definition
MIBK	Methyl Isobutyl Ketone
CAGR	Compound Annual Growth Rate
DAA	Diacetone alcohol
MO	Mesityl oxide
DIBK	Diiasobutyl ketone
M	Mesitylene
IPA	Isopropyl alcohol
W	Water
FCI	Fixed Capital Investment
TCI	Total capital investment
ROI	Return on investment
DCFRR	Discounted Cash Flow Rate of Return

References

1. Kurbatova, M.V.; Cherentsova, M.I.; Raskulova, T.V.; Fereferov, M.Y.; Ryabtsov, A.Y. Low-temperature dewaxing of oils in the presence of an individual solvent. *Vestn. ANSTU* **2018**, *12*, 69–72.
2. Smirnova, A.; Grigor'eva, L.; Ostroukhov, N. Extraction of water-soluble phenols from shale-chemical process water. *Solid Fuel Chem.* **2016**, *50*, 371–375. [CrossRef]
3. Roy, C.B.; Awual, M.; Goto, M. Effect of inorganic salts on ternary equilibrium data of propionic acid-water-solvents systems. *J. Appl. Sci.* **2007**, *7*, 1053. [CrossRef]
4. Zhiyong, Z.; Wei, Q.; Weiyang, F.; Yigui, L. A study on stoichiometry of complexes of tributyl phosphate and methyl isobutyl ketone with lithium in the presence of $FeCl_3$. *Chin. J. Chem. Eng.* **2012**, *20*, 36–39.
5. Xiang, W.; Liang, S.; Zhou, Z.; Qin, W.; Fei, W. Lithium recovery from salt lake brine by counter-current extraction using tributyl phosphate/$FeCl_3$ in methyl isobutyl ketone. *Hydrometallurgy* **2017**, *171*, 27–32. [CrossRef]
6. Verma, R.K. *Single-Step MIBK Process*; IHS Markit: London, UK, 2019.

7. Grosse, Y.; Baan, R.; Secretan-Lauby, B.; El Ghissassi, F.; Bouvard, V.; Benbrahim-Tallaa, L.; Guha, N.; Islami, F.; Galichet, L.; Straif, K. *Carcinogenicity of Chemicals in Industrial and Consumer Products, Food Contaminants and Flavourings, and Water Chlorination Byproducts*; Elsevier: Amsterdam, The Netherlands, 2011.
8. Expert Market Research. "Global Methyl Isobutyl Ketone Market Report and Forecast 2022–2027." Methyl Isobutyl Ketone Market Report, Size, Share, Analysis 2022–2027, 2021. Available online: https://www.expertmarketresearch.com/reports/methyl-isobutyl-ketone-market (accessed on 23 August 2022).
9. Intelligence Mordor. "Methyl Isobutyl Ketone (MIBK) Market Size, Share: 2022–2027: Industry Analysis." Methyl Isobutyl Ketone (MIBK) Market Size, Share | 2022–2027 | Industry Analysis, 2021. Available online: https://www.mordorintelligence.com/industry-reports/methyl-isobutyl-ketone-market (accessed on 23 August 2022).
10. Rutala, A.W.; Weber, D.J. *Guideline for Disinfection and Sterilization in Healthcare Facilities, 2008*; CDC: Atlanta, GA, USA, 2008.
11. Weissermel, K.; Arpe, H.-J. *Industrial Organic Chemistry*; John Wiley & Sons: Hoboken, NJ, USA, 2008.
12. Winter, F.; van Dillen, A.J.; de Jong, K.P. Single-stage liquid-phase synthesis of methyl isobutyl ketone under mild conditions. *J. Mol. Catal. A Chem.* **2004**, *219*, 273–281. [CrossRef]
13. Torres, G.; Apesteguia, C.R.; di Cosimo, J.I. One-step methyl isobutyl ketone (MIBK) synthesis from 2-propanol: Catalyst and reaction condition optimization. *Appl. Catal. A Gen.* **2007**, *317*, 161–170. [CrossRef]
14. Lei, Z.; Li, J.; Li, C.; Chen, B. Improvement of separation process of synthesizing MIBK by the isopropanol one-step method. *Korean J. Chem. Eng.* **2006**, *23*, 264–270. [CrossRef]
15. Cosimo, J.; Torres, G.; Apesteguia, C. One-step MIBK synthesis: A new process from 2-propanol. *J. Catal.* **2002**, *208*, 114–123. [CrossRef]
16. Mattos, V.L.; Noronha, F.B.; Monteiro, J.L.F. Bifunctional metal/base catalysts (Pt/X) for the direct synthesis of MIBK from acetone. *J. Catal.* **2002**, *209*, 166–176. [CrossRef]
17. Mayevskiy, M.; Frolkova, A.; Frolkova, A. Separation and purification of methyl isobutyl ketone from acetone + isopropanol + water + methyl isobutyl ketone + methyl isobutyl carbinol + diisobutyl ketone mixture. *ACS Omega* **2020**, *5*, 25365–25370. [CrossRef] [PubMed]
18. Bagabas, A.A.; Mokhtar, M.; Akhmedov, V.; Narasimharao, K.; Basahel, S.N.; Al-Rabiah, A. Ru–C–ZnO Composite Catalysts for the Synthesis of Methyl Isobutyl Ketone via Single Step Gas Phase Acetone Self-Condensation. *Catal. Lett.* **2014**, *144*, 1278–1288. [CrossRef]
19. Gamman, J.J.; Jackson, S.D.; Wigzell, F.A. Synthesis of methyl isobutyl ketone over Pd/MgO/SiO_2. *Ind. Eng. Chem. Res.* **2010**, *49*, 8439–8443. [CrossRef]
20. Bagabas, A.A.; Akhmedov, V.M.; Al-Rabiah, A.; Mostafa, M.M.M. Synthesizing and Utilizing Novel Nano Crystalline Zinc Chromate Supported Nano Palladium Catalyst. U.S. Patent US 7,951,976, 31 May 2011.
21. Kelkar, V.V.; Vaibhav, V.; Lionel, K.; O'Young, C.; Wibowo, M.; Kelkar, H.; Chung, C. Two-Step System and Method for the Production of Methyl Isobutyl Ketone. U.S. Patent US 9120734B2, September 2015.
22. Shylesh, S.; Hanna, D.; Gomes, J.; Canlas, C.G.; Head-Gordon, M.; Bell, A.T. The Role of Hydroxyl Group Acidity on the Activity of Silica-Supported Secondary Amines for the Self-Condensation of n-Butanal. *ChemSusChem* **2015**, *8*, 466–472. [CrossRef] [PubMed]
23. Talwalkar, S.; Mahajani, S. Synthesis of methyl isobutyl ketone from acetone over metal-doped ion exchange resin catalyst. *Appl. Catal. A Gen.* **2006**, *302*, 140–148. [CrossRef]
24. Nicol, W.; du Toit, E.L. One-step methyl isobutyl ketone synthesis from acetone and hydrogen using Amberlyst®CH28. *Chem. Eng. Process. Process Intensif.* **2004**, *43*, 1539–1545. [CrossRef]
25. Zhang, S.; Wu, P.; Yang, L.; Zhou, Y.; Zhong, H. An efficient bifunctional catalyst of TiO_2 coating and supported Pd on cordierite for one-pot synthesis of MIBK from acetone. *Catal. Commun.* **2015**, *71*, 61–64. [CrossRef]
26. Rao, P.V.R.; Kumar, V.P.; Rao, G.S.; Chary, K.V. Vapor phase selective hydrogenation of acetone to methyl isobutyl ketone (MIBK) over Ni/CeO_2 catalysts. *Catal. Sci. Technol.* **2012**, *2*, 1665–1673. [CrossRef]
27. Nikolopoulos, A.; Jang, B.-L.; Spivey, J. Acetone condensation and selective hydrogenation to MIBK on Pd and Pt hydrotalcite-derived MgAl mixed oxide catalysts. *Appl. Catal. A Gen.* **2005**, *296*, 128–136. [CrossRef]
28. Chen, Y.; Hwang, C.; Liaw, C. One-step synthesis of methyl isobutyl ketone from acetone with calcined Mg/Al hydrotalcite-supported palladium or nickel catalysts. *Appl. Catal. A Gen.* **1998**, *169*, 207–214. [CrossRef]
29. Ho, C.R.; Zheng, S.; Shylesh, S.; Bell, A.T. The mechanism and kinetics of methyl isobutyl ketone synthesis from acetone over ion-exchanged hydroxyapatite. *J. Catal.* **2018**, *365*, 174–183. [CrossRef]
30. Schmitt, K.; Disteldorf, J.; Flakus, W.; Hubel, W. Process for preparing methyl isobutyl ketone and catalyst. U.S. Patent US 3,953,517, April 1976.
31. Rase, H.F. *Handbook of Commercial Catalysts: Heterogeneous Catalysts*; CRC Press: Boca Raton, FL, USA, 2000.
32. Kozhevnikova, F.E.; Kozhevnikov, I.V. One-step synthesis of methyl isobutyl ketone from acetone catalysed by Pd supported on ZnII–CrIII mixed oxide. *J. Catal.* **2006**, *238*, 286–292. [CrossRef]
33. Guisnet, M. Heterogeneous catalysis and fine chemicals. In *International Symposium on Heterogeneous Catalysis and Fine Chemicals (1988: Poitiers, France)*; Distributors for the US and Canada; Elsevier Science Pub. Co.: Amsterdam, The Netherlands, 1988.
34. Zhu, Q.; Duan, H.; Lin, B.; Zhu, Y.; Hu, Y.; Zhou, Y. Higher Acetone Conversion Obtained Over a TiO2–Pd Bifunctional Catalyst for Liquid-Phase Synthesis of Methyl Isobutyl Ketone: The Role of Al_2O_3 Support. *Catal. Lett.* **2019**, *149*, 2636–2644. [CrossRef]

35. Lin, B.; Xu, F.; Hu, Y.; Du, X.; Zou, Y.; Xie, H.; Wang, K.; Zhou, Y. An efficient multifunctional catalyst for one-pot synthesis of methyl isobutyl ketone: Phosphor-doped h-BN with adjustable acid-base property as support. *Catal. Commun.* **2021**, *150*, 106276. [CrossRef]
36. Duan, H.; Wang, Z.; Cui, L.; Lin, B.; Zhou, Y. Stability Investigation of a Supported TiO_2–Pd Bifunctional Catalyst over the One-Pot Liquid-Phase Synthesis of Methyl Isobutyl Ketone from Acetone and H_2. *Ind. Eng. Chem. Res.* **2018**, *57*, 12358–12366. [CrossRef]
37. Zhu, Q.; Dai, H. Exposure Assessment of Emerging Chemicals and Novel Screening Strategies. In *Emerging Chemicals and Human Health*; Springer: Berlin, Germany, 2019; pp. 9–26.
38. Zhu, Y.; Lin, B.; Hu, Y.; Cai, Z.; Xie, H.; Wang, K.; Zhou, Y. Enhanced stability of Pd/SPS catalyst over the one-pot liquid-phase synthesis of methyl isobutyl ketone by adding GO. *Mol. Catal.* **2019**, *478*, 110609. [CrossRef]
39. Liu, M.; Liu, G.; Zhou, Y.; Han, K.; Ye, H. Sulfonated graphene oxide supported Pd bifunctional catalyst for one-pot synthesis of methyl isobutyl ketone from acetone with high conversion and selectivity. *J. Mol. Catal. A Chem.* **2015**, *408*, 85–90. [CrossRef]
40. Al-Rabiah, A.A.; Bagabas, A.A.; Malik, A.V. Low Pressure One-Step Gas-Phase Process for Production of Methyl Isobutyl Ketone. U.S. Patent US 8,101,805, 24 January 2012.
41. Al-Rabiah, A.A.; Bagabas, A.A.; Akhmedov, V.M. Low Pressure One-Step Gas-Phase Process for Production of Methyl Isobutyl Ketone. E.P. Patent E.P. 2,532,642 B1, 25 January 2017.
42. Peters, M.; Timmerhaus, K.; West, R. *Plant Design and Economics for Chemical Engineers*, 5th ed.; McGraw Hill: New York, NY, USA, 2016.
43. Alexandre, C.D. Chapter 19—Economic Evaluation of Projects. In *Computer Aided Chemical Engineering*; Elsevier: Amsterdam, The Netherlands, 2014; Volume 35, pp. 717–755.
44. Thomas, M. Chemical Engineering Plant Cost Index. In *Chemical Engineering*; The U.S. Department of Labor's Bureau of Labor Statistics: Washington, DC, USA, 2022.
45. Chemanalyst. Chemical Price Analysis, Chemical Latest Prices | ChemAnalyst. 2022. Available online: https://www.chemanalyst.com/Pricing/Pricingoverview (accessed on 23 August 2022).
46. Turton, R.; Bailie, R.C.; Whiting, W.B.; Shaeiwitz, J.A. *Analysis, Synthesis, and Design of Chemical Processes*, 5th ed.; Prentice Hall: Hoboken, NJ, USA, 2018.
47. Hong, S.C.; Musso, T.J. *Simons Oil-Price Shocks and the Chemical Industry: Preparing for a Volatile Environment*; McKinsey and Company: Atlanta, GA, USA, 2015.
48. Knight, P. *Oil and Gas Market Volatile, Petrochemical Market May Follow Suit If Invasion Prolongs*; University of California Press: Berkley, CA, USA, 2022.

processes

Article

Pipeline Two-Phase Flow Pressure Drop Algorithm for Multiple Inclinations

Andrés Cepeda-Vega [1,2,*], Rafael Amaya-Gómez [1,*], Miguel Asuaje [3,4], Carlos Torres [5], Carlos Valencia [2] and Nicolás Ratkovich [2]

1 Chemical Engineering Department, Universidad de los Andes, Cra 1 Este No 19A-40, Bogotá 111711, Colombia
2 Industrial Engineering Department, Universidad de los Andes, Cra 1 Este No 19A-40, Bogotá 111711, Colombia; cf.valencia@uniandes.edu.co (C.V.); n.rios262@uniandes.edu.co (N.R.)
3 Department of Energy Conversion and Transport, Universidad Simón Bolivar, Cra 59 No 59-65, Caracas 1086, Venezuela; asuajem@usb.ve
4 Frontera Energy, Cll 100 No 9-25, Bogotá 110221, Colombia
5 Thermal Science Department, Universidad de los Andes, Mérida 5101, Venezuela; ctorres@ula.ve
* Correspondence: af.cepeda10@uniandes.edu.co (A.C.-V.); r.amaya29@uniandes.edu.co (R.A.-G.)

Abstract: A Generalized Additive Model (GAM) is proposed to predict the pressure drop in a gas–liquid two-phase flow at horizontal, vertical, and inclined pipes based on 21 different dimensionless numbers. It is fitted from 4605 points, considering a fluid pattern classification as Annular, Bubbly, Intermittent, and Segregated. The GAM non-parametric method reached high prediction capacity and allowed a great degree of interpretability (i.e., it helped to visualize and test statistical inference), considering that each predictor's marginal effects could be described, unlike in other Machine Learning (ML) methods. The prediction capacity of the GAM model for the pressure gradient obtained an adjusted R^2 and a mean relative error of 99.1% and 12.93%, respectively. This capacity is maintained even when ignoring Bubbly flow in the training sample. A regularization technique to filter some variables was used, but most of the predictors must maintain the model's high predictive ability. For example, dimensionless numbers such as the Reynolds, Froude, and Weber numbers show p-values of less than 0.01% to explain the pressure gradient in the different flow patterns. The model performs adequately on 500 randomly sampled data points not used to fit the model with an error lower than 15%. The variable importance for the model and the relationship with the pressure gradient is evaluated based on the obtained splines and p-values.

Keywords: pressure gradient; gas–liquid two-phase flow; flow patterns; Generalized Additive Model; dimensionless numbers

Citation: Cepeda-Vega, A.; Amaya-Gómez, R.; Asuaje, M.; Torres, C.; Valencia, C.; Ratkovich, N. Pipeline Two-Phase Flow Pressure Drop Algorithm for Multiple Inclinations. *Processes* **2022**, *10*, 1009. https://doi.org/10.3390/pr10051009

Academic Editor: Jean-Claude Assaf

Received: 2 March 2022
Accepted: 18 March 2022
Published: 19 May 2022

Publisher's Note: MDPI stays neutral with regard to jurisdictional claims in published maps and institutional affiliations.

1. Introduction

Transporting fluids in the Oil and Gas Industry, for instance, from the well to the processing plant (upstream sector), represents a challenge. This challenge requires a complex configuration of pipes that contain multiphase mixtures of gases and liquids at different inclinations and operating conditions. The design of this two-phase configuration includes essential parameters for operators such as pressure gradient or void fraction, which are avoided in single-phase flows, in which reliable methods to calculate frictional pressure drop are available (e.g., the Moody chart or Colebrook-White equation). For two-phase flow, there are complex dynamics between the gas and liquid within the pipe that makes these predictions much more challenging [1].

Several approaches estimate or predict the frictional pressure drop in two-phase flows based on empirical correlations such as the ones proposed by Hagedorn and Brown [2] or Beggs and Brill [3]. These correlations depend strongly on the data used and the conditions under which they were built; therefore, their application range is restricted. For example,

Hagedorn and Brown's [2] original correlations' were not based on conservation and momentum balances, although the development of the original paper's correlation followed some physical phenomena present in the system. The mechanistic modeling approaches for different pipe configurations were developed later using mass and momentum balances. Overall, most of these models depend on how flow patterns are classified, allowing assumptions about the geometry of the cross-sectional area distribution of the phases, which are necessary to study their interfacial properties. For instance, Xiao, Shonham, and Brill [4] proposed a pressure drop model for Annular, Bubbly, Intermittent, and Stratified flow patterns using the continuity and momentum equations by assuming uniform liquid height in the film zone, which they stated is good enough for most practical applications.

Despite these efforts, mechanistic models such as those by Hasan and Kabir, and Ansari, among others still need to select from various closure laws to work them properly given to their limitations to apply different two-phase flow correlations [5]. Mechanistic models cannot explain some phenomena, so some empirical correlations need to be introduced, and since these closure laws are empirically based, they undermine the objective of mechanistic modeling [5]. An alternative approach uses drift flux models [6], which consider the mixture as a whole rather than two separated phases [7]. In this direction, drift flux models follow a more rigorous approximation than the two-phase mechanistic models; however, they assume homogeneous flow, which can be a strong assumption when the difference of density between fluids is significantly enough like in the case of oil-water systems [8]. Moreover, there are different available empirical flow pattern transition maps based on the phases' superficial velocities (e.g., Taitel and Dukler [9]), which depend on the flow being near horizontal or vertical. These maps help visualize the most likely flow pattern of the mixture and are essential for any mechanistic model that aims to describe the two-phase flow dynamics. These maps have been improved considering that statistical and machine learning tools such as Neural Networks (NN) and kernels have been considered from a probabilistic-based approach. Overall, Machine Learning (ML) implements data with statistical learning methods and algorithms to "learn" based on some predefined labeled (supervised) or unlabeled (unsupervised) features. Within the supervised approaches they are classification or regression models, whereas in the case of unsupervised learning algorithms for clustering or dimensionality reduction. In this regard, data recollection plays a relevant role to develop these kind of approaches as in work done in the Tulsa Unified Fluid Flow Project (TUFFP) by Zhang et al. [10]. For instance, Amaya-Gómez et al. [11] dealt with the flow pattern classification problem using k-nearest neighbors and a Bayesian approach based on a relevant experimental database. There are other examples that have implemented Machine Learning approaches for two-phase or multiphase related to pattern flow recognition, flow dynamics, liquid hold-up calculation, including Refs. [5,12–15].

Hybrid approaches combining a probabilistic or a machine learning method with a mechanistic model can help to obtain better estimations of both void fraction and pressure gradient. For example, the probabilistic method proposed by Amaya-Gómez et al. [11] can be implemented jointly with the TUFFP model [10]. The flow pattern classification could be handled with the probabilistic approach, while the momentum and continuity equations would be employed to calculate the pressure gradient. Other alternatives include predicting the flow patterns and pressure gradients based on linear regression or probabilistic methods. For example, Kanin et al. [5] used Random Forest, Gradient Boosting Decision Trees, Support Vector Machines, and Artificial Neural Networks to predict flow patterns, void fraction, and pressure gradients in a three-stage model. Compared to the mechanistic models proposed by Ansari et al. [16], Xiao, Shonham, and Brill [4], and the TUFFP Unified [10,17], the results reported lower prediction errors. Another interesting example is reported by Hernández et al. [18], where the authors use decision trees to identify which phenomenologically based correlations are best to predict in specific flow patterns. The approach of Hernández et al. [18], which selects closure relationships or even different ML models, can also be improved by optimization tools; for example, Mohammadi, Papa,

Pereyra, and Sarica [19] show that using a genetic algorithm can improve a mechanistic model 277% by selecting closure relationships.

Although these statistical and machine learning methods may have good predictions, they also lack in-depth interpretability, considering that most of these methods follow a black-box approach that may not provide deep insights into the available data. Consequently, this study seeks to predict the total pressure drop in gas–liquid two-phase flow using statistical methods that facilitate the results' interpretation. Here, the interpretation is associated with the possibility of evaluating p-value tests, displaying predictors contributions, and estimating predictors' confidence intervals. These features allow identifying the risk built in the model. For example, confidence intervals can be used to see where the splines effects are dubious because of the lack of observations. Furthermore, p-values can tell which predictors are statistically insignificant and should not be considered. Following these requirements, Generalized Additive Models and Regression Splines would be considered. This strategy retains interpretability and statistical inference capability while having a functional prediction capacity when it is continuous and smooth [20]. Generalized Additive Models (GAM) are a middle ground tool for prediction and can be used to generalize general linear models. GAMs are a tool for both prediction and statistical inference based on a standard generalized model with a link function consisting of a mix of linear or parametric functions of the covariates and other non-parametric non-linear functions. They have been implemented in different fields such as agriculture and medicine. See Hastie and Tibshirani [21] for some applications. Overall, a GAM can be used in any case where linear or logistic regression fits the objective's scope. GAM can find non-linear relationships between predictors and the response in both the classification and regression settings in addition to the statistical background of linear or logistic methods. The predictors correspond with different well-known dimensionless numbers and the response with the pressure gradient for this work.

Previous Uses of Dimensionles Numbers

In multiphase flow literature, numerous cases implemented dimensionless numbers as inputs for correlation or mechanistic models. Their use is attractive because available databases usually have a limited range of reported substances. Hence, considering correlations based on known dimensionless quantities allows the model to be expanded for different substances if there are sufficient records of the corresponding dimensionless number. One of the most famous models in the two-phase flow, where dimensionless numbers are contemplated, is the Taitel and Dukler [9] mechanistic model for near horizontal inclinations. They manipulate the momentum and mass balance equations mathematically to express them on known dimensionless quantities known as the Lockhart–Martinelli parameter and a dimensionless hydrostatic pressure drop for stratified flow. The correlation works by solving the equations that depend on these dimensionless numbers to obtain the liquid film height, which is then used with geometrical assumptions to find the rest of the interest variables.

The use of dimensionless numbers in mechanistic models is still prevalent. For example, the Ansari, Sylvester, Shoham, and Brill [16] models used the Lockhart–Martinelli parameter, the two-phase mixture Reynolds number, and the friction factors based on the development's superficial velocities of their model. The mechanistic model of Petalas and Aziz [22] used the Reynolds number of both the gas and liquid phases based on the hydraulic diameters, the Froude number of the liquid phase, and other dimensionless quantities, considering the relationship with parameters to be found, such as liquid film height or wetted perimeter. The model of Zhang et al. [10] implemented dimensionless numbers, both in determining the unified flow model and supporting closure laws. Namely, the closure laws they used for the liquid entrainment in the gas core are based on the gas Weber, Froude, and both liquid and gas Reynolds. Furthermore, they depend on the dimensionless ratio of the densities and viscosities of the phases.

Recent statistical approaches such as Artificial Neural Networks (ANN) or ensembled decision trees also have considered dimensionless numbers as inputs since they improve predictive ability and generalizability for validation. For example, Ozbayoglu and Ozbayoglu [23] published an ANN model to predict frictional pressure losses and flow patterns based on the superficial Reynolds number of both phases as inputs. Other examples include the machine learning approach proposed by Kanin et al. [5] which implements dimensionless features such as the Reynolds and Froude numbers and additional proposed dimensionless quantities. These features train multiple Machine Learning (ML) algorithms, which they cross-compare and obtain high R^2 values in their predictions.

This article is organized as follows: Section 2 presents the theoretical framework of the GAM and how flow patterns are mainly classified. Section 3 describes the implemented database and the main dimensionless numbers used as predictors. Section 4 presents the proposed methodology. Section 5 summarizes the essential findings using the GAM model. Finally, Section 6 shows some concluding remarks.

2. Theoretical Basis

2.1. Generalized Additive Models (GAM)

In the case of regression with a continuous response (normal distribution), the model with k covariates, $j-1$ of them with linear effects and $k-j$ with non-linear forms, has the following general expression:

$$Y = \beta_0 + \beta_1 x_1 + \cdots + \beta_{j-1} x_{j-1} + f_j(x_j) + \cdots + f_k(x_k) + \epsilon \quad (1)$$

where functions $f_j(x_j)$ may describe flexible relations estimated non-parametrically, and ϵ is the model's error, representing the intrinsic random variability. Generalized Additive Models are no different from generalized linear models (e.g., logistic regression) except for non-parametric parts of the function. However, instead of assuming a linear relationship between predictors and linked response, some smooth function is found that represents well the effect of the non-parametric terms on the variable to be predicted. Please note that additivity implies that each of the independent variables has a functional effect on the response Y, and the addition of the terms composes the joint multivariate function. The resulting model has the advantage of being interpretable in the effects of each covariate. This happens because each function is the central effect of the respective variable on the response. When the function is not additive, the high order interactions hide the interpretation, given that margins cannot separate the multivariate function. Please note that interpretability in this model means that each covariate's effect on the response may be separated; however, if a margin is non-parametrically estimated, there is no analytical expression of the function in general. Moreover, the estimators are defined pointwise and are easier to describe graphically. Another sense in which interpretability is advantageous is that the model can be used to construct p-values and confidence intervals. These can be used to assess the risk of bias and overfitting in the model. In this sense, the modeler can know which covariates have a high degree of confidence in their estimates and if the pressure gradient can be explained more accurately.

The general function in Equation (1) allows including flexible dependence of the response on the covariates. Some smoothness restrictions must be imposed to control that each marginal function truly reflects the relation's central tendency and is not too influenced by the random error ϵ. This process is called regularization and avoids over-fitting to the data. It does not mean that the resulting estimator is necessarily a smooth function, but that the function's total variation is controlled. As a result, another advantage is that the model, in general, has better predictive performance if it is used to predict the mean response on data points that are out of the sample used to estimate the model, which is a consequence of the non-linear nature of the relationship with the response. Simultaneously, the additive form imposes a structural restriction on the mean multivariate function that avoids the deterioration of the estimations when the number of variables increases, a characteristic of non-parametric methods usually referred to as the curse of dimensionality.

However, this flexibility comes at the cost of finding how to represent the non-parametric functions in some way and choosing how "smooth" they should be [24]. The problem of representation is generally solved by using a finite number of basis functions for each function $f_j(x_j)$. The following equation defines the non-parametric term as:

$$f_j(x_j) = \sum_{b=1}^{B} b_{j_b}(x_j)\beta_{j_b} \tag{2}$$

where $b_{j_b}(x_j)$ is the b-th base function for the covariate x_j and β_{j_b} are their corresponding coefficients. The most straightforward approach is to use a polynomial basis; nevertheless, this approach has the problem implied by the Taylor theorem, which states that the basis will be a good fit in the neighborhood of a single specified point but not a good representation of the global function. A convenient approach to solve this problem is to use basic functions that are more local, such as a piecewise function defined by a certain number of knots. Figure 1 shows an example of a piecewise linear function with two knots (spline basis with degree 1). A spline basis is a generalization of this idea; in each interval, a polynomial function is defined, requiring that continuity and smoothness are imposed on the knots. With this type of basis, the estimated function's flexibility may be controlled by the number of knots and each piecewise polynomial degree.

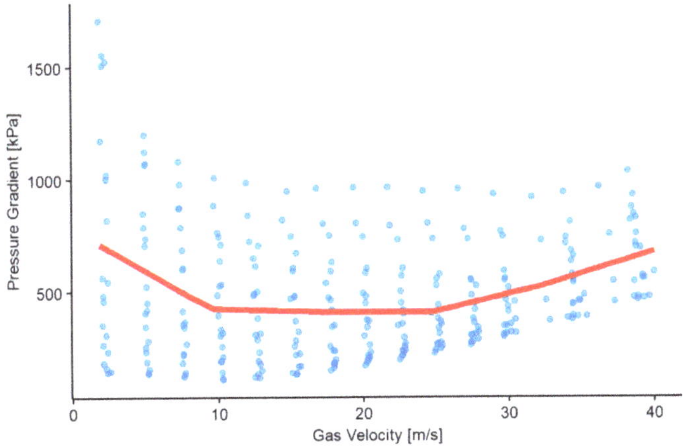

Figure 1. Piecewise linear regression fit on the regression of Gas Velocity to predict Pressure Gradient. Using the data of Alsaadi [25].

A more convenient alternative for calibrating the GAM, i.e., selecting the appropriate degree of flexibility in each non-parametric function, considers a penalization term on the estimation by ordinary least squares, on which $f_j(x_j)$ is limited in order to avoid too much variation (wiggliness). A common penalization is the squared second derivative, for which the estimation $(\hat{\beta}_0, \ldots, \hat{\beta}_{j-1}, \hat{f}_j, \ldots, \hat{f}_k)$ results of minimizing:

$$\sum_{i=1}^{n}\left(y_i - \beta_0 - \cdots - \beta_{j-1}x_{j-1,i} - f_j(x_{ji}) - f_k(x_{ki})\right)^2 + \sum_{h=j}^{k}\lambda_h \int f_h''(x_h)^2 dx_h \tag{3}$$

where y_i and x_{ji} represents the i-th observation of the response variable and the j-th covariate, respectively. The parameters λ_h must be calibrated to use the optimal degree of flexibility.

Since the smoothness parameters of the splines are unknown at the start of the algorithm, they are initialized, assuming they do not affect the response. The algorithm cycles find the smoothing parameter (λ_h) of each of the non-parametric terms until they reach a stable value. This algorithm is known as backfitting [21]. The convergence of the method is not only fast but guaranteed, with the only restriction that the splines for each covariate x_j are centered [26]:

$$\sum_{i=1}^{n} f_j(x_{ji}) = 0 \quad (4)$$

This condition implies that the splines do not represent the variable's marginal effects on the response but on the centered response (the mean is subtracted). This situation influences the interpretation of the statistical inference results but not the model's predictive capability. The primary advantage of GAM models is interpretability and hypothesis testing. For example, the following hypothesis can be considered:

Ho: *Variable X is in the model.*

Ha: *Variable X is not in the model.*

This can be done, as in linear regression, using the F partial test as follows [20]:

$$F = \frac{\frac{SSE_{reduced} - SSE_{full}}{df_{reduced} - df_{full}}}{MSE_{full}} \quad (5)$$

where SSE and MSE represent the sum and mean of the reduced and the entire model's square residuals; moreover, df represents the degrees of freedom of each model (the number of parameters used in it). The F statistic can be interpreted as any statistical test. If it has a higher value than a specified significance level in the corresponding accumulated density distribution, then the null hypothesis can be rejected. There can also be a hypothesis made on the nature of the parameter, as shown below:

Ho: *Variable X has a linear effect.*

Ha: *Variable X has a nonparametric effect.*

These tools are a definite advantage that improves the model's interpretability compared to other machine learning approaches such as Neural Networks or ensembled trees.

2.2. Flow Patterns Considered

Flow pattern classification can be ambiguous, depending on the methodology of each author. For example, Aggour [27] classified patterns between annular (froth and mist), bubble (froth), Slug, Slug Froth, and transitions between slug-bubble and slug-annular. Brito [28] classified them between Annular, Bubble, Intermittent Flow, which combines Slug Flow and Elongated Bubble, and Stratified Wavy. In the database used, each author has their way of naming each flow pattern. Therefore, to have a consistent classification and broader categories, the categories shown in Table 1 are considered.

The classification proposed in Table 1 is consistent with previous works in the literature such as those by Amaya-Gómez et al. [11], Brito [28], Fan [29], and Mantilla [30]. Furthermore, transitions between flow patterns can be seen, especially from intermittent to annular and intermittent to bubble. Figure 2 shows the flow patterns used in this document and illustrates their characteristics. It can be seen in Figure 2 that bubbly flow refers to small spherical bubbles of gas, which are dispersed in the liquid phase. Intermittent flow is comprised of large and elongated bubbles of gas that form liquid pistons of considerable length. Stratified flow is a typical flow with a well-defined interface between the fluids is formed for any cross-sectional cut of the pipe. Stratified Flow disappeared for an inclination greater than 70° (upward direction). Annular flow occurs at high void fractions and generally is comprised of a high-velocity gas phase that pushes the liquid phase outwards.

Table 1. Classification of flow patterns for prediction.

	Flow Pattern Considered			
	Annular Flow (A)	**Bubbly Flow (BF)**	**Intermittent Flow (IT)**	**Segregated Flow (SG)**
Original description	-Annular -Annular/Intermittent -Annular Froth -Annular Mist -Annular Wavy -Churn Annular	-Bubble -Bubble/Intermittent -Bubble Froth	-Intermittent/Annular -Intermittent/Bubble -Slug Churn -Slug Flow -Slug Froth	-Stratified -Stratified Wavy

Figure 2. Illustrative shape of the flow patterns used in this work. (**a**) Bubbly Flow. (**b**) Intermittent Flow. (**c**) Segregated Flow. (**d**) Annular Flow.

3. Materials: Database and Dimensionless Numbers

3.1. Database Presentation

This work implements the data reported at the TUFFP database, including 43,565 records, out of which 5011 registered both a flow pattern and total pressure drop. There are 4605 cases of pipes inclined upwards and 406 cases of pipes inclined downwards. A negative ΔP sign implies the pipe gains pressure instead of losing it because of a drop in the potential energy. The database used in the development of the proposed model consists of these 5011 points; these records are described in more detail in Table 2.

This database comprises 18 references, out of which some of them are commonly used to validate different mechanistic models and correlations experimentally, such as the data of Andritsos [31] being used to validate the model of Zhang et al. [17]. Table 2 shows the maximum and minimum values for the primary measurements of the database, including the diameter, length-to-diameter (L/D), superficial velocities (V_L^S and V_G^S), the angle of inclination of the pipe (θ), the flow pattern classifications they contain, temperature, pressure, and pressure gradient. From these variables, the last one is to be predicted. Flow patterns are classified in the database as 22.31% annular, 6.41% bubbly, 44.12% Intermittent, and 27.16% segregated.

Figure 3 presents the distribution of the liquids and gases used in the database. There are more substances in the liquid phase, but around two thirds of the records used water or glycerin dilutions. The gas phase is composed mainly of air and should not be generalizable for lighter or denser gases. This problem can be solved using dimensionless numbers, as will be described in this work.

Table 2. Overview of the data of each author in the database. Diameter, L/D ratio, superficial velocities of the phases, angle of inclination, flow patterns, temperature, pressure, and pressure gradient are presented.

Author	Data	D [mm]	L/D	V_L^S [m/s]	V_G^S [m/s]	θ [°]	FP	T [°C]	P [kPa]	$\Delta P/L$ [kPa/m]
Aggour [27]	412	11.7	130	0.3139–10.58	0.07–96	90	A, BF, IT	20–39	106–493	0.176–68.52
Alsaadi [25]	288	76.2	232	0.0099–0.101	1.83–40	2 to 30	A, IT, SG	20	100	0.11–1.71
Andritsos [31]	535	25.1–95.3	105–616	0.0003–0.335	0.8–98.9	0	SG	10–27	98–196	0–2.23
Asante [32]	476	25.4–76.2	499–1496	0.0001–0.2	15–30	0	A, SG	15	270	0.001–0.04
Brito [28]	346	50.8	374	0.0097–2.961	0.09–7.7	0	A, BF, IT, SG	21–50	102–211	0.013–7.24
Chung et al. [33]	159	50.8	447	0.0476–0.704	0.28–7.8	−90	A, SG	21–49	109–232	−4.419–0.7
Fan [29]	321	50.8–149.6	335–377	0.0003–0.052	4.93–25.7	−2 to 2	A, IT, SG	15–17	101–222	0.005–0.72
Gockal [34]	183	50.4	375	0.01–1.76	0.09–20.3	0	A, IT, SG	21–38	101	0.033–18.49
Güler-Quadir [35]	102	88.9	8099	0.0366–1.667	0.12–17.7	90	A, BF, IT	20	102	0.45–4.25
Karami [36]	106	152.4	370	0.01–0.027	7.4–22.6	0	SG	22	163	0.011–0.1
Kokal [37]	870	25.8–76.3	313–931	0.03–3.048	0.03–19.4	−9 to 9	IT	11–31	227–365	−1.101–9.62
Kouba [38]	52	76.2	418	0.1524–2.137	0.3–7.4	0	IT	14–42	196–386	0.001–0.03
Mantilla [30]	142	50.8–152.4	53–253	0.0033–0.1	1.5–82.2	0	A, SG	20	104–224	0–6.26
Mora and Zegrí	39	21	228.6	0.0481–1.623	0.22–10.6	0	A, BF, IT, SG	20	103–271	0.006–3.29
Rezkallah [39]	380	11.7	130	0.0219–10.577	0.05–129.6	90	A, BF, IT	19–31	96–403	−0.158–62.7
Skopich [40]	35	50.8–101.6	150–300	0.009–0.05	7.99–29.5	90	A	29–37	104–129	0.176–1.48
Sujumnong [41]	232	11.7	130	0.0457–8.46	0.03–118.4	90	A, BF, IT	17–32	101–343	0.475–62.89
Vijay [42]	333	11.7	130	0.0122–10.607	0.05–140.3	90	A, BF, IT	20–31	99–513	0.276–90.58
Minimum		11.7	53	0.0001	0.03	−90		10	96	−4.419
Maximum		152.4	8099	10.607	140.3	90		50	513	90.58

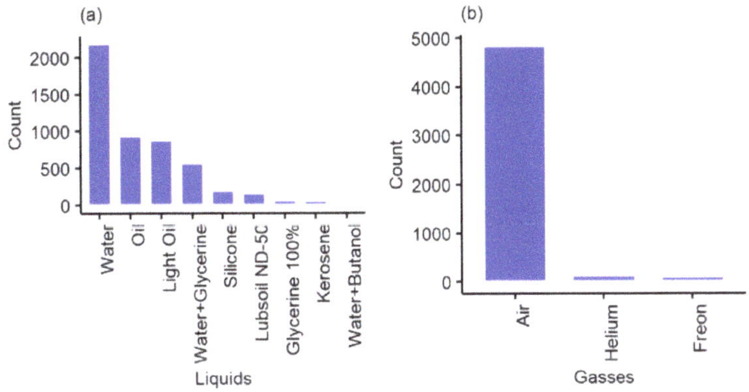

Figure 3. Distribution of phases in the database, (**a**) Liquids, and (**b**) Gases.

Figure 4 shows the distribution of the pipes' diameter, L/D relation, superficial velocities, temperature, and pressure of the pipe. It can be seen in Figure 4 that the database cases have temperature readings in conditions near ambient temperature. This is a long, vertically inclined, water-air pipe. Regarding the pressure, most of the records reported near-standard pressure, but there are also pressurized pipes.

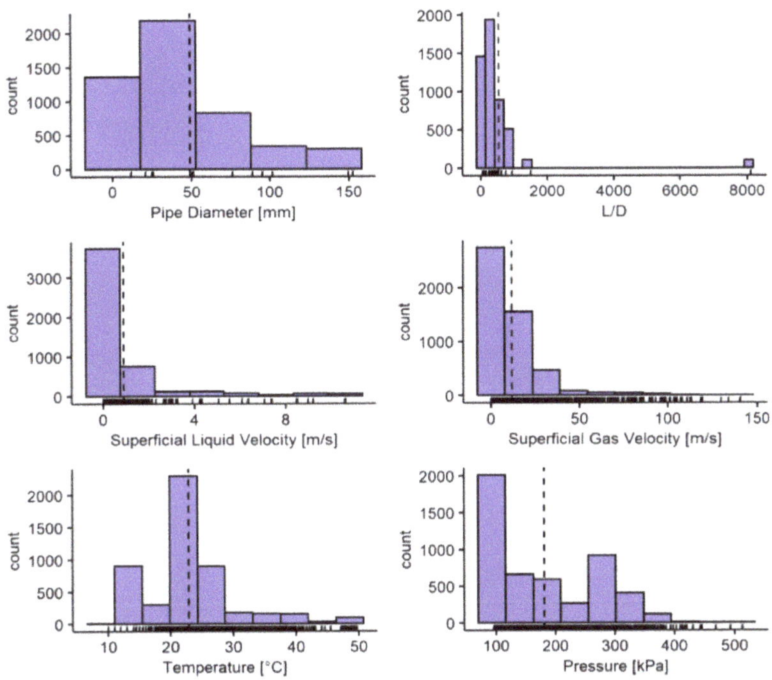

Figure 4. Distributions of essential variables in the database. The dotted line corresponds with the sample's mean.

Figure 5 depicts the pipe inclinations, obtaining that 29.81% of the records had vertical upward pipes (i.e., 90°) and 3.2% vertical downward pipes (i.e., −90°), 44.3% horizontal pipes, and 22.7% inclined pipe cases. More downward cases should be included to improve the database. Figure 5 shows there must be precautions to generalize the final model as a tool for predicting pressure drop in any condition, especially considering the inclination distribution and the overrepresentation of water and air mixtures in the database. However, dimensionless numbers are the chosen strategy to draw broader conclusions from the model proposed in this work.

Table 3. Dimensionless numbers considered in this work as predictors.

Dimensionless Number	Symbol	Equation		Relationship Explained
Gas Quality	w_G	$\frac{\dot{m}_G}{\dot{m}_G + \dot{m}_L}$	(6)	Gas mass flow rate to total flow rate
Gas Reynolds	Re_G	$\frac{\rho_G\, V_G^S\, D}{\mu_G}$	(7)	Inertial forces to viscous forces
Liquid Reynolds	Re_L	$\frac{\rho_L\, V_L^S\, D}{\mu_L}$	(8)	
Mixture Reynolds	Re_M	$\frac{\rho_M\, V_M^S\, D}{\mu_M}$	(9)	
Gas Weber	We_G	$\frac{\rho_G\, {V_G^S}^2\, D}{\sigma}$	(10)	Inertial forces to surface tension
Liquid Weber	We_L	$\frac{\rho_L\, {V_L^S}^2\, D}{\sigma}$	(11)	
Mixture Weber	We_M	$\frac{\rho_M\, {V_M^S}^2\, D}{\sigma}$	(12)	
Gas Froude	Fr_G	$\frac{V_G^S}{\sqrt{g\,D}}$	(13)	Intertial forces to gravity
Liquid Froude	Fr_L	$\frac{V_L^S}{\sqrt{g\,D}}$	(14)	
Mixture Froude	Fr_M	$\frac{V_M^S}{\sqrt{g\,D}}$	(15)	
Gas friction factor	f_G	$C\, Re_G^{-n}$	(16)	Friction factors as in single-phase flow *.
Liquid friction factor	f_L	$C\, Re_L^{-n}$	(17)	
Interfacial friction factor	f_I	$f_G\left(1 + 14.3\sqrt{1-w_G}\left(\sqrt{\frac{V_G^S\,\rho_G}{\rho_{0G}}}\right)\right)$	(18)	Interfacial friction factor approximation for segregated/annular flow.
Froude Rate	F_t	$\left(\frac{w_G^3(\dot{m}_G + \dot{m}_L)^2}{\rho_G^2\, g\, D(1-w_G)A^2}\right)^{0.5}$	(19)	Vapor Kinetic Energy to the energy required to lift the liquid
Y Tailel Dukler Number	Y_{TD}	$\frac{2D(\Delta\rho)g\sin(\theta_{rad})}{f_G\,\rho_G\,{V_G^S}^2}$	(20)	Hydrostatic pressure loss to gas frictional pressure loss

Table 3. *Cont.*

Dimensionless Number	Symbol	Equation		Relationship Explained
Eötvös	E_o	$\frac{g\,D^2(\rho_L-\rho_G)}{\sigma}$	(21)	Gravitational to capillary forces.
Lockhart Martinelli	χ	$\frac{f_L\,\rho_L\,V_L^{S^2}}{f_G\,\rho_G\,V_G^{S^2}}$	(22)	Frictional liquid loss to Frictional gas loss.
Viscous number	N_μ	$\frac{V_M^S\,\mu_L}{g\,D^2\,\Delta\rho}$	(23)	Viscous forces to gravitational forces
Length Diameter Ratio	L/D	$\frac{L}{D}$	(24)	Pipe geometry
Dimensionless Temperature	$\tilde{T}$	$\frac{T}{T_0}$	(25)	
Dimensionless Pressure	$\tilde{P}$	$\frac{2P_o}{f_G\,\rho_G\,V_G^{S^2}}$	(26)	The initial pressure of pipe to gas pressure loss.

* C is 16, and n is 1 for $Re < 3000$ (Laminar). C is 0.046, and n is 0.2 for $Re \geq 3000$ (Turbulent). As used by Taitel and Dukler.

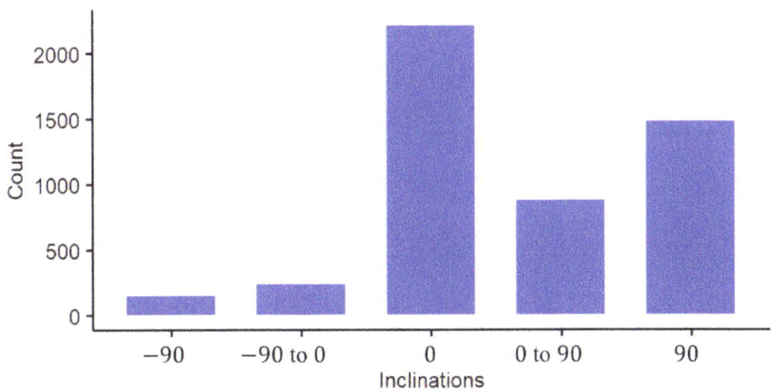

Figure 5. Distribution of the inclinations of the pipe in the database.

3.2. Dimensionless Numbers

This work aims to propose a sufficiently robust pressure drop predictive model, seeking to use less training data to be commercially viable. Therefore, it is essential to use known dimensionless numbers as predictors in the database, such as those presented in Table 3.

In Table 3, there are present some mixture properties. All of them are defined the same way; for example, mixture density is defined as:

$$\rho_M = w_G\rho_G + (1 - w_G)\rho_L \tag{27}$$

Every mixture property is defined as in the last equation except mixture velocity because it is traditionally defined in two-phase flow literature as:

$$V_M^S = V_G^S + V_L^S \tag{28}$$

3.3. Notes on the Dimensionless Numbers Selected

There are multiple ways to define dimensionless numbers; mechanistic models for two-phase flow can also be enriched using geometrical parameters such as the hydraulic diameter or wetted perimeter. Since this work follows a statistical-based model, the dimensionless numbers selected contemplated only those based on easily measured variables. In this regard, geometrical considerations were not considered. Some considerations of the dimensionless shown in Table 3 are described below:

- Gas Quality w_g serves as a proxy for the void fraction of the model. While the void fraction can give more information about the two-phase flow system, it would require a two-step model to be calculated. This two-step model can be contemplated as a future improvement of the proposed model.
- The Reynolds Re, Weber We, and Froude Fr numbers are defined using superficial velocities and the nominal diameter. Although this is a simple definition for two-phase flow, the statistical model can find non-linear relationships and does not need to have as detailed information as in a mechanistic model. To include both phases' information, the superficial velocities of the liquid, phase, and mixture are contemplated (Equation (28)). Therefore, three different dimensionless numbers (liquid, gas, and mixture) are based on the velocity used. Multiple mechanistic models and statistical models have used these numbers to calculate two-phase flow dynamics.
- The friction factors of the liquid and gas phases are defined as reported by Taitel and Dukler [9]. The interfacial friction factor expression uses the expression used by Zhang et al. [10], which is a closure law of that mechanistic model.
- The Froude rate F_t has been used in different studies to calculate void fraction. Graham et al. [43] obtained a correlation for the void fraction using the Froude rate and the Lockhart–Martinelli parameter χ in the stratified flow pattern and nearly horizontal pipes. The idea behind their inclusion is to incorporate further information about the void fraction.
- The Y Taitel–Dukler number was used in Taitel's and colleagues' 1976 correlation Taitel and Dukler [9]; they found this grouping was relevant for nearly horizontal gas–liquid flow. The Lockhart–Martinelli parameter also appears as a relevant dimensionless grouping in the same study. The idea to adimensionalize the loss of the hydrostatic pressure based on the gas pressure loss is used in this document when defining the dimensionless pressure $\tilde{P}$.
- The Eötvös number E_0 is relevant for small diameter pipes because it captures the system's capillary forces well. The dominance of the surface tension can help to predict flow pattern maps where superficial tension dominates. Brauner and Moalem [44] found surface tension maps and used the Eötvös number to define where superficial tension dominated as part of their analysis.
- The viscous number N_μ is a dimensionless grouping used by Al-Safran et al. [45] to develop an empirical model that tries to correlate liquid hold-up using different dimensionless groupings for two-phase slug flow.
- The length to diameter ratio and dimensionless temperature are used since there is information available in the database used in this model.

4. Methodology

The distribution of the dimensionless numbers throughout the database is smoother than the non-processed parameters. Furthermore, the ranges of applicability of the model can be interpreted in the ranges of the dimensionless numbers, which substantially improves the model's generalizability. Figure 6 shows the histograms of some of these dimensionless numbers.

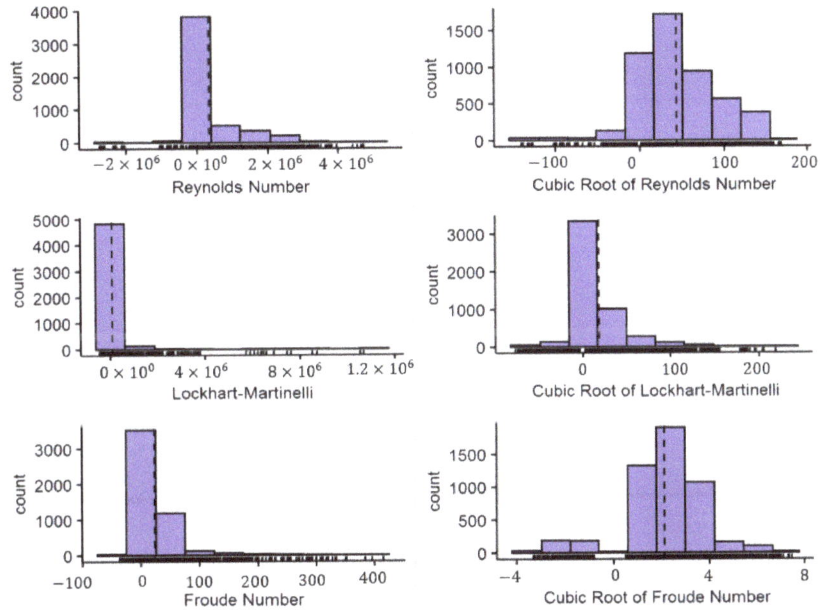

Figure 6. Distribution of some of the dimensionless numbers presented in Table 3. Note: Reynolds and Froude are calculated for the liquid phase.

Figure 7a shows the distribution of the response variable (i.e., the pressure gradient). This histogram is left-skewed (e.g., the mean is left to the main peak), so using the correct transformation can improve the normality of the data, which for GAM is not a requirement, but it still can help improve the prediction. The proposed transformation is a cubic root as depicted in Equation (29), which can be used even for negative numbers, unlike a logarithmic transformation. The obtained transformation for the variable response is shown in Figure 7b. Please note that it improves the shape of the histograms. Most of the dimensionless numbers used have similar histogram distributions. As shown in Figure 7b, the cubic root transformation centers the data distribution and reduces skewness. This transformation helps the regression method.

$$Y = \sqrt[3]{\frac{\Delta P}{L}} \tag{29}$$

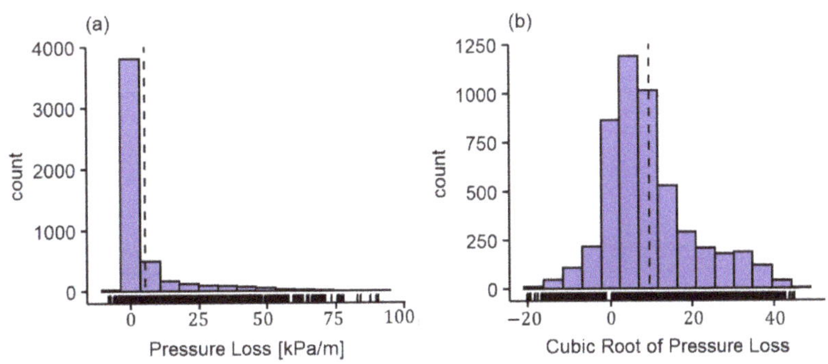

Figure 7. Distributions of Response Variable for Regression. (**a**) Histogram of pressure loss. (**b**) Histogram of transformed response.

Additionally, most collinear predictors (i.e., more correlated variables) were removed to reduce the set of predictors from Table 3 using the findCorrelations function in the caret package in R-3.6.3. For this purpose, a correlation threshold of 0.85 high was used to prevent valuable information from being removed from the model. Predictors with a higher correlation than the threshold were the gas Reynolds and the mixture Froude, so they were discarded from the model.

The model assumes that all splines are non-parametric thin-plate splines, yet since these splines are penalized, some of them can end having linear relations with the response. In the final model, to be precise, 19 dimensionless numbers were used, so the model was comprised of 57 non-parametric different covariates since each dimensionless number and flow pattern interaction is considered and since Bubbly flow was merged with the Intermittent flow. The proposed methodology is shown in Figure 8, following a data collection, model training, and using the model to predict the pressure drop.

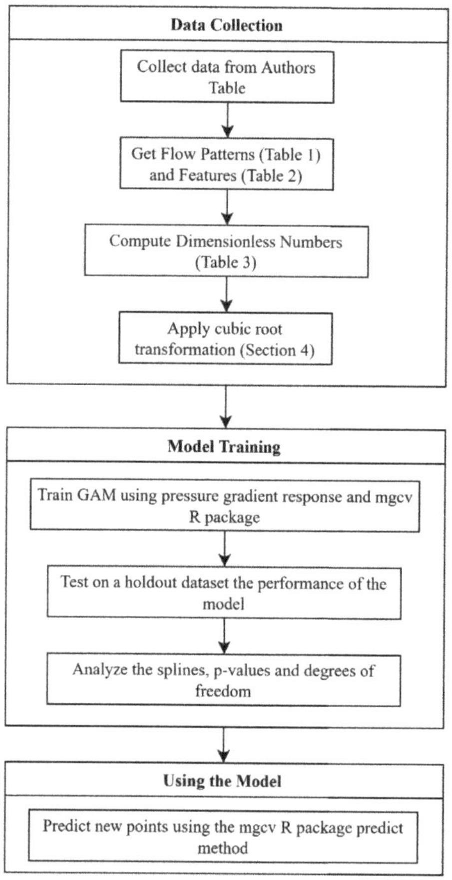

Figure 8. Scheme of the Proposed methodology.

The GAM was fitted using the mgcv library in R, built by Simon Wood [24]. It has been used in multiple applications in different fields [24]. For each of the covariates with non-linear relation, the basis representations used were thin-plate regression splines that are low-rank smoothers, meaning that they have fewer coefficients than data to smooth. The main idea of this approach is to use a penalized model as the one represented in Equation (3), on which the resulting basis representation is in the form of splines with knots placed on each of the observed points xi. The mgcv library uses different computational approaches to fit the splines in a fast and efficient way. As described in Section 2.1, the calibration of smoothing parameters is performed by generalized cross-validation (GCV). This method is very close to leave-one-out cross-validation (LOOCV), on which the prediction performance is measured on each data point that is left out of the sample. LOOCV fits the model on all the points except one, then evaluates the model performance on that point. Then, LOOCV takes each point in the sample and repeats the procedure, averaging each out-of-sample point's mean square error. GCV is a less computationally demanding but accurate approximation.

5. Results and Discussion

5.1. Predictive Capability

There are multiple ways to calibrate a GAM; both the predictors and the complexity of each spline must be found in a way the model performs better on a test dataset. Section 1 shows that the splines could be calibrated using the back-fitting algorithm and the thin-plate spline definition. Marra and Wood [46] conducted a study on variable selection. The conclusion was that shrinkage approaches perform significantly better than competing methods in terms of predicting ability. The best-performing methods of variable selection were the double penalty and the eigenvalue shrinkage approaches. Because of the more straightforward interpretability of the double penalty approach, it is considered for this study. In either case, both approaches were the best methods for variable selection, just as compared by Marra and Wood [46].

The double penalty method works using all the predictors from Table 3 except the ones removed due to their high correlations under a regularization method such that fewer essential predictors are taken out of the model. The double penalty method is similar to LASSO (Least Absolute Shrinkage and Selection Operator) variable selection in which the optimization objective function is changed so that models with fewer variables are preferred. It is a double penalty method because both the thin-plate splines and the objective function are penalized. Some splines are taken out of the model, and the ones staying are not too complex.

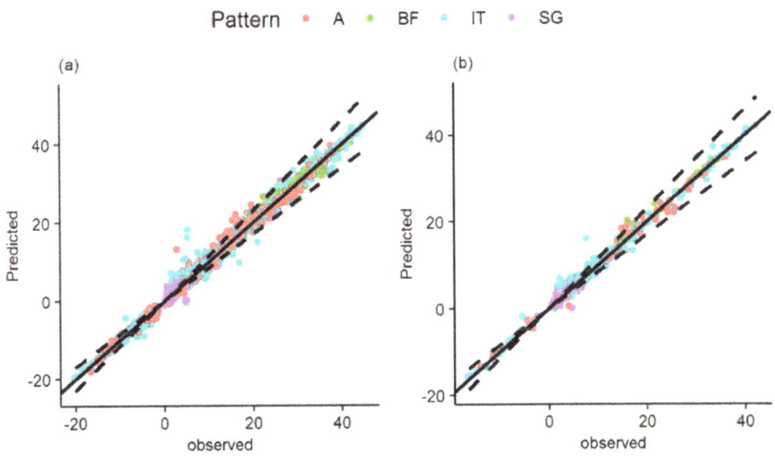

Figure 9. Results of the best model using double penalty regularization and all the dimensionless numbers proposed. A: Annular, BF: Bubbly Flow, IT: Intermittent, SG: Segregated. (**a**): Observed vs. Predicted plot on all the datasets with 15% error lines. (**b**): Observed vs. Predicted plot on a 500-points sample taken as a validation set, with 15% error lines

Figure 9a shows the results obtained by the GAM using all the predictors with the double penalty approach. Figure 9b shows the case where the GAM is calibrated with all the data except for 500 points randomly sampled from the original data set, then validating the results on these points such that the predictive capability of the model is shown for new data. Please note that the model follows closely the mean and is contained in the 15% error lines on downward (negative) and upward (positive) cases.

The relative error (RE) is defined as the absolute difference between the response and predicted values over the original or observed response. The mean relative error (MRE) for each flow pattern will then be used to illustrate the model's predictive capability (Equation (30)) where m_j is the total number of observations in the corresponding flow pattern.

$$MRE = \sum_{i=1}^{m_j} \frac{|Y^i_{obs} - Y^i_{pred}|}{m_j \, Y^i_{obs}}, \quad \forall j \in \{A, BF, IT, SG\} \tag{30}$$

Furthermore, for model comparison, AIC, or the Akaike information criterion based on the model's likelihood function, was considered. The measure tries quantifying a trade-off between goodness of fit and the number of parameters in the model, penalizing overfitting. Then the AIC can be compared between models, where a smaller value means better quality.

This relative error is similar to models such as the one proposed by Zhang et al. [10] or correlations such as those by Andritsos, Williams, and Hanratty [47]. The total deviance explained by the model is 99.2%, and the adjusted R^2 is 99.1%, which both mean the model already explains the total variance practically in the data and that it will perform adequately in a test sample, as shown both in Figure 9 and Table 4. The distribution of standardized residuals shown in Figure 10 has no identifiable patterns, and the histogram suggests normality in the residuals, indicating that the model is correctly specified. It has a light skewness to the left, meaning the model can sometimes underpredict the pressure drop. However, this is on a small subset of the whole sample and should not be a concern considering the model's predictive capability. Furthermore, the model performs adequately both in downward and upward cases. Residuals which are used in Figure 10a,b are defined using the standardized residuals expression:

$$Residuals = \frac{Y^i_{obs} - Y^i_{pred}}{\sqrt{\dfrac{\sum_{i=1}^{n} Y^i_{obs} - Y^i_{pred}}{n-1}}} \tag{31}$$

Table 4. MSE and relative error results on the train and test dataset used for the GAM calibrated with the shrinkage approach.

Set	Pattern	MSE	Mean Relative Error (%)
Train	Total	0.9239	11.720
	A	12.404	11.860
	BF	0.1661	1.330
	IT	12.919	10.390
	SG	0.2474	16.290
Test	Total	17.473	11.080
	A	41.118	9.660
	BF	0.2860	2.340
	IT	16.122	10.960
	SG	0.3680	14.020
Deviance Exp.	99.20%	AIC	13,110.21

A: Annular, BF: Bubbly Flow, IT: Intermittent, SG: Segregated.

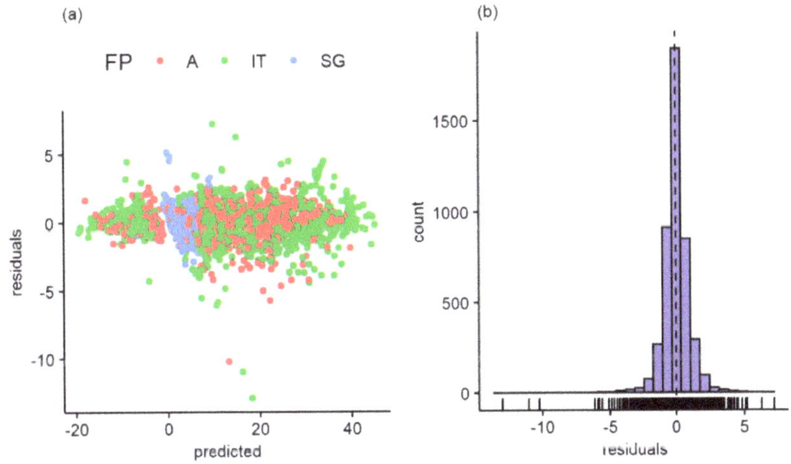

Figure 10. Distribution of residuals of GAM with double penalty regularization. A: Annular, BF: Bubbly Flow, IT: Intermittent, SG: Segregated. (**a**) Predicted vs. residuals scatterplot. The discontinuity between the two clusters in this plot is the division between upward and downward cases. (**b**) Histogram of residuals.

This model's critical weakness is using most of the predictors, although some of them are removed considering the tool proposed by Marra and Wood [46] for a total of 19 dimensionless numbers as predictors in the GAM with a double penalty. Considering the four different flow patterns, 76 different splines need to be calibrated for the trained model. This situation makes interpretability difficult because it is required to review all the significant predictors' responses. Since the appeal of the GAM is a compromise between predictive capacity and interpretability, this is a weakness that should be addressed.

The first approach to reduce the number of features in the model considered analyzing the information in Table 4, specifically, the relative error in each flow pattern. This table suggests that the model is beneficial to predict the bubbly flow. Therefore, the proposed trade-off would reduce the model's capability to predict bubbly flow and reduce the number of features, which should improve the interpretability and confidence interval of each spline. For this purpose, the model was trained to ignore bubbly flow cases, assuming they were intermittent flow cases since intermittent flow is the most prevalent in the database. This assumption means that the model does not distinguish between bubbly flow and intermittent flow on a dataset. The hypothesis is that if the model can predict the pressure drop in intermittent flow cases correctly, then it will continue to predict the bubbly flow cases correctly in a test dataset. This approach's advantage was the reduction from 76 initial predictors to 57 predictors (i.e., 19 are immediately removed) before the double penalty regularization is done since a whole flow pattern was removed from the training process. The results of the model without bubbly flow are shown in Table 5.

Table 5 shows that a GAM that predicts correctly intermittent flow can still have excellent bubbly flow performance. The AIC increased only by 3.3%, but only 50 splines remained to be interpreted after double penalty regularization, which is statically relevant with a 5% significance level. Nevertheless, 50 splines are still much information to consider, but the model could significantly reduce predictive capacity seeking to prune more. The problem with this model is that it starts predicting less accurately segregated flow pattern cases, as shown in Table 5, where the MSE of the segregated flow is the smallest in this case.

Table 5. Results on train and test set for the GAM without bubbly flow training.

Set	Pattern	MSE	Mean Relative Error (%)
Train	Total	0.983	12.347
	A	10.590	11.028
	BF	1.335	3.579
	IT	12.584	9.221
	SG	0.4062	19.269
Test	Total	1.237	12.926
	A	12.971	9.420
	BF	13.553	4.413
	IT	15.585	10.557
	SG	0.6087	19.977
Deviance Exp.	99.20%	AIC	13,562.47

A: Annular, BF: Bubbly Flow, IT: Intermittent, SG: Segregated.

Another way to improve each flow pattern's relative error is by calibrating a GAM for each flow pattern using all the dimensionless numbers and the double penalty approach. The problem with this approach is that it is more sensible to flow pattern misclassification since the splines are not built based on the other flow patterns but only on the application. Furthermore, since 4% relative error is still acceptable for bubbly flow, only three models were constructed, and the intermittent flow model was used to predict bubbly flow cases. However, the results of the three models combined on the same train and test datasets are the following:

As shown in Table 6, using a different model for each flow pattern has similar results to the best model found. However, as seen in the MSE of the annular flow pattern, it can have worse results on a test data set since it has fewer data to train each model. Furthermore, statistical metrics such as AIC are not comparable to previous models since the training samples are different.

Table 6. Results on train and test set by using three models, one for annular, one for stratified, and one for intermittent.

Set	Pattern	MSE	Mean Relative Error (%)
Train	A	0.9160	9.842
	BF	13.517	3.593
	IT	12.828	9.288
	SG	0.2522	17.401
Test	A	67.762	13.042
	BF	13.400	4.246
	IT	15.485	10.389
	SG	0.3411	17.623

A: Annular, BF: Bubbly Flow, IT: Intermittent, SG: Segregated.

5.2. Results of the GAM

This section presents the results that do not distinguish between the intermittent and bubbly flow. The results of the full GAM model with all the flow patterns are shown in the annexes. The results of the methodology that fitted one GAM for each flow pattern are also presented. The statistical significance results of the GAM can be seen in Tables 7 and 8.

Table 7. Parametric coefficients of the GAM. The base case is the annular flow pattern, and the bubbly flow is mixed with intermittent flow.

	Parametric Coefficients			
	Estimate	**Std. Error**	***t* Statistic**	***p*-Value**
Intercept (A)	6.433	13.185	0.488	0.626
IT	3.749	13.191	0.284	0.776
SG	−9.939	31.325	−0.317	0.751

A: Annular, IT: Intermittent, SG: Segregated.

Table 8. Results of the non-parametric tests of the GAM. Estimated degrees of freedom is proportional to the complexity of each spline.

Approximate Significance of Smooth Terms:									
Spline	**E. dof**	**R. dof**	**F Statistic**	***p*-Value**	**Spline**	**E. dof**	**R. dof**	**F Statistic**	***p*-Value**
$s(w_G)$:A	0.975	9	4.346	<0.0001	$s(f_I)$:A	4.082	9	11.453	<0.0001
$s(w_G)$:IT	3.286	9	2.531	<0.0001	$s(f_I)$:IT	3.730	9	20.569	<0.0001
$s(w_G)$:SG	0.000	9	0.000	0.4610	$s(f_I)$:SG	2.798	9	1.648	0.0001
$s(Re_L)$:A	5.499	19	3.505	<0.0001	$s(Ft)$:A	5.797	9	5.18	<0.0001
$s(Re_L)$:IT	14.410	19	10.184	<0.0001	$s(Ft)$:IT	5.089	9	5.532	<0.0001
$s(Re_L)$:SG	0.897	16	0.543	0.0009	$s(Ft)$:SG	0.301	9	0.048	0.1159
$s(Re_M)$:A	10.790	18	23.061	<0.0001	$s(Y_{TD})$:A	10.100	17	10.531	<0.0001
$s(Re_M)$:IT	14.220	19	12.224	<0.0001	$s(Y_{TD})$:IT	15.630	19	85.988	<0.0001
$s(Re_M)$:SG	10.740	19	5.336	<0.0001	$s(Y_{TD})$:SG	1.028	7	7.34	<0.0001
$s(We_G)$:A	12.710	19	6.471	<0.0001	$s(Eo)$:A	3.483	5	7.002	<0.0001
$s(We_G)$:IT	3.498	19	1.065	<0.0001	$s(Eo)$:IT	4.953	5	150.699	<0.0001
$s(We_G)$:SG	13.680	18	26.050	<0.0001	$s(Eo)$:SG	3.967	5	122.387	<0.0001
$s(We_L)$:A	7.581	18	3.293	<0.0001	$s(\chi)$:A	0.939	19	0.803	<0.0001
$s(We_L)$:IT	10.290	19	4.777	<0.0001	$s(\chi)$:IT	8.757	18	4.911	<0.0001
$s(We_L)$:SG	0.000	10	0.000	0.5311	$s(\chi)$:SG	0.000	19	0	0.2248
$s(We_M)$:A	6.757	9	11.820	<0.0001	$s(N_\mu)$:A	15.690	19	32.64	<0.0001
$s(We_M)$:IT	4.515	9	6.304	<0.0001	$s(N_\mu)$:IT	12.700	19	101.09	<0.0001
$s(We_M)$:SG	1.294	8	0.535	0.0149	$s(N_\mu)$:SG	6.823	17	1.929	<0.0001
$s(Fr_G)$:A	4.958	19	1.166	<0.0001	$s(L/D)$:A	4.719	6	14.115	<0.0001
$s(Fr_G)$:IT	13.680	19	13.242	<0.0001	$s(L/D)$:IT	5.775	6	127.099	<0.0001
$s(Fr_G)$:SG	14.750	19	21.040	<0.0001	$s(L/D)$:SG	5.413	6	61.614	<0.0001
$s(Fr_L)$:A	14.550	19	13.200	<0.0001	$s(\tilde{T})$:A	6.231	9	9.122	<0.0001
$s(Fr_L)$:IT	12.270	19	48.426	<0.0001	$s(\tilde{T})$:IT	8.404	9	13.493	<0.0001
$s(Fr_L)$:SG	4.008	15	1.655	<0.0001	$s(\tilde{T})$:SG	6.269	9	51.724	<0.0001
$s(f_G)$:A	0.671	5	0.407	0.0533	$s(\tilde{P})$:A	8.509	15	11.022	<0.0001
$s(f_G)$:IT	9.217	19	7.514	<0.0001	$s(\tilde{P})$:IT	17.080	19	9.897	<0.0001
$s(f_G)$:SG	4.554	10	9.246	<0.0001	$s(\tilde{P})$:SG	9.002	17	11.042	<0.0001
$s(f_L)$:A	0.000	16	0.000	0.5305					
$s(f_L)$:IT	0.001	15	0.000	0.4237					
$s(f_L)$:SG	1.419	19	0.213	0.0450					
R-sq (adj)	99.10%	Deviance Explained		99.10%	Scale Estimated	10.738			

A: Annular, IT: Intermittent, SG: Segregated. E.dof: Estimated degrees of freedom of each spline. R.dof: Reference degrees of freedom for each spline.

Tables 7 and 8 show that the parametric terms only comprise the flow pattern, which is a categorical variable. The baseline level is the annular flow pattern; it only changes how the model is interpreted but not the results in any way. The flow patterns are not significant based on their *p*-values on the response's mean but are significant as they change the splines' shape. In Table 8, the need to use a different spline for each flow pattern is revealed since each spline's estimated degrees of freedom are different for the same dimensionless number. The reference degrees of freedom are the maximum the splines can have. The back-fitting algorithm optimizes the best-estimated degrees of freedom in this range that perform better in theory based on metrics such as generalized cross-validation.

The reference degrees of freedom are arbitrarily selected, but they should be large enough to prevent under smoothing, considering that too large degrees of freedom may have computational costs. The *p*-values are based on the cumulative F distribution, and they are the probability accumulated in the right tail. If this probability is low, it means that the term has a statistically significant non-linear effect on the pressure gradient. In this model, most non-parametric terms significantly affect the response, even after the double penalty approach. So, filtering the variables more could reduce the performance of the model significantly. As can be seen, the adjusted R^2 is the same as the deviance, which is a clear indication that most of the variables are necessary to explain the variance in the model. They are also both near the unity and show that the model predicts the pressure gradient.

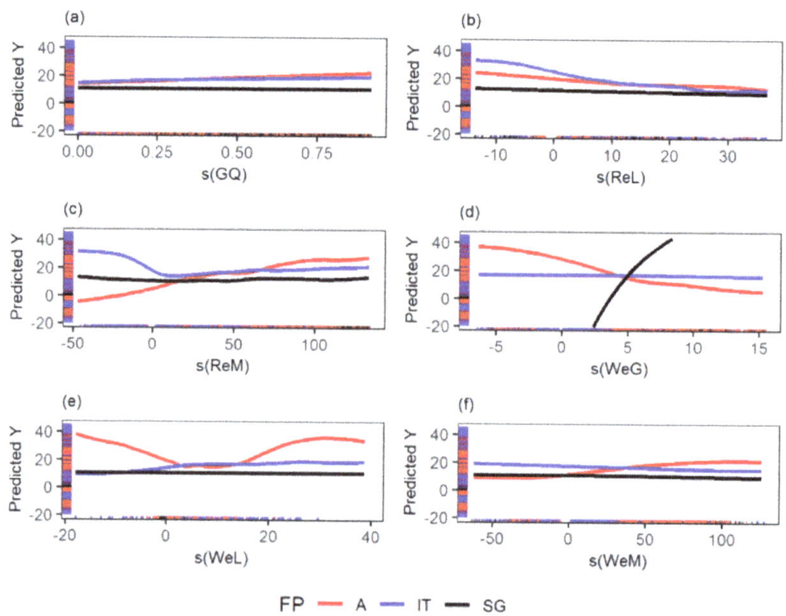

Figure 11. Splines of the cube root of the dimensionless numbers against the pressure gradient's centered cube root. A: Annular, IT: Intermittent, SG: Segregated. (**a**): Gas Quality. (**b**): Liquid Reynolds. (**c**): Mixture Reynolds. (**d**): Gas Weber. (**e**): Liquid Weber (**f**): Gas Weber.

Figures 11–13 show the splines of the model. These figures could provide attractive insights since each dimensionless effect on the pressure drop can be compared, considering flow patterns. Figure 11 shows some of the insights that can be drawn from the GAM methodology. For example, it is known that liquid entrainment can have an impact on stratified flow. The gas weber number is related to how liquid droplets can enter the gas phase and start a transition between flow patterns. For example, Zhang, Wang, Sarica, and Brill [10] used a liquid entrainment empirical correlation for stratified flow. This can explain the strong dependence on the gas weber number for the stratified case, as shown in that spline, comparing it to intermittent and annular flow.

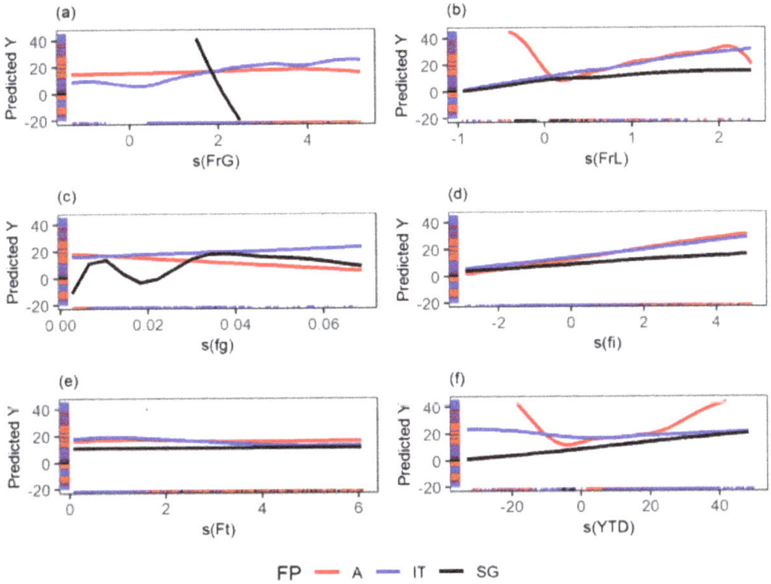

Figure 12. Splines of the cube root of the dimensionless numbers against the pressure gradient's centered cube root. A: Annular, IT: Intermittent, SG: Segregated. (**a**): Gas Froude. (**b**): Liquid Froude. (**c**): Gas friction factor. (**d**): Interfacial friction factor. (**e**): Froude Rate. (**f**): Y Taitel and Dukler parameter.

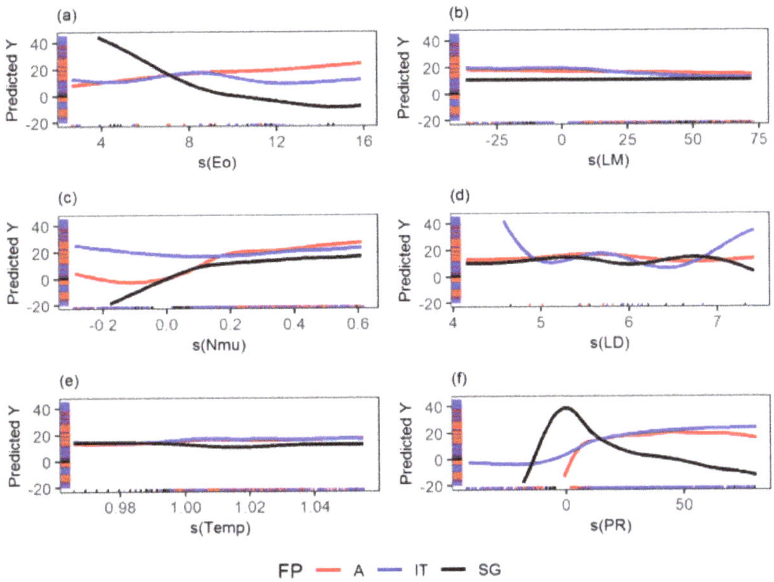

Figure 13. Splines of the cube root of the dimensionless numbers against the pressure gradient's centered cube root. A: Annular, IT: Intermittent, SG: Segregated. (**a**): Eötvös Number. (**b**): Lockhart–Martinelli parameter. (**c**): Viscous number. (**d**): Length diameter ratio. (**e**): Dimensionless Temperature. (**f**): Dimensionless Pressure.

The splines shown in Figures 11–13 demonstrate the relationship between the cube root of the dimensionless predictors and the cube root of the pressure gradient. The splines are plotted in a window such that they comprise the mean of the dimensionless predictor and two standard deviations. These plots also show the model's limitations since points outside the showed window should not be predicted with confidence. Nonetheless, all the dimensionless numbers have a large window for prediction.

In some cases, the relations are as expected, such as in the liquid Reynolds, which is inversely proportional to the pressure gradient in annular and intermittent flow. Similar to the single-phase flow case, friction losses decrease in a turbulent flow. However, mixture Reynolds shows the inverse relationship in annular, which can be explained by the gas phase's drag experienced or the gas's compressibility, and because the annular flow has the more considerable void fraction typically.

5.3. Performance on the Authors' Datasets

The predicted vs. observed plot of the model on the data of the most prevalent author of each flow pattern is showed: Rezkallah [39] for annular flow; Aggour [27] for bubbly flow; Kokal [37] for intermittent flow, and Andritsos [31] for stratified flow.

Figure 14 displays the excellent agreement of the model to the authors' data. Again, stratified flow is the one with the worst performance. It was shown that fitting different GAMs for each flow pattern could solve this problem. Furthermore, stratified flow is the one that can be easier to predict because of the well-defined liquid height so that a hybrid model approach could be used—for example, using a mechanistic model for stratified flow and the GAM for the other flow patterns.

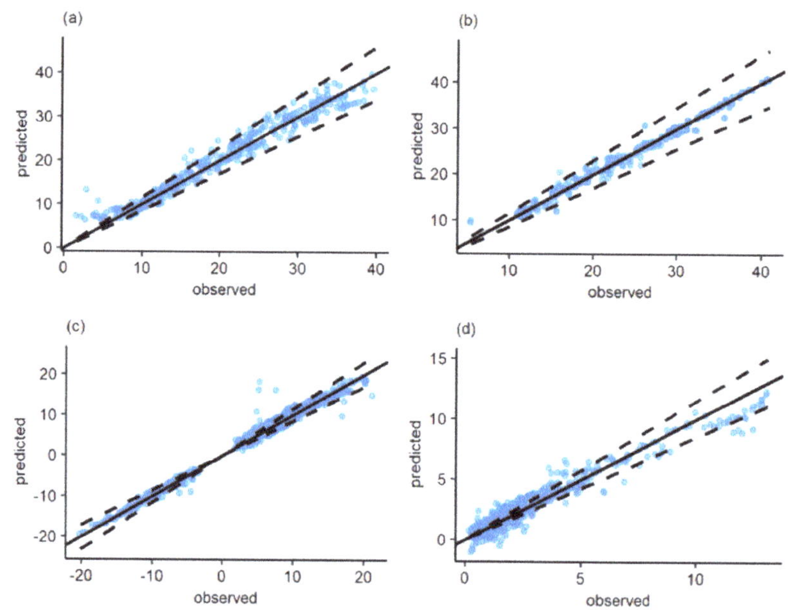

Figure 14. Observed vs. predicted plot with 15% error lines for the four authors mentioned. (**a**): Rezkallah [39] (6.868% relative error). (**b**): Aggour [27] (2.691% relative error). (**c**): Kokal [37] (13.05% relative error). (**d**): Andritsos [31] (17.04% relative error).

5.4. Performance against Mechanistic and Black-Box Models

After describing the proposed approach's prediction capabilities, it was compared in terms of predictive performance against both mechanistic and statistical-machine learning tools. In this regard, the Tulsa Unified Model proposed by Zhang et al. [10], which is a mechanistic model for all types of flow patterns and inclinations, and a gradient boosting model (GBM) approach, which is an application of the work done by Kanin et al. [5] on the database described in Section 3, which is a machine learning model with high predictive capacity were compared with the proposed GAM. Figure 15 shows the mean relative error and mean square error of the three approaches for each dataset recollected

To calibrate Kanin et al.'s [5] GBM, the R xgboost library was used. It was calibrated by margins following the same sequence as in the original work. Two GBM were built, one using the same predictors as the authors of the model and using the same dimensionless numbers used in this model. The one with more predictors, which uses the dimensionless numbers of this work, is the one with lower test-MSE, so that result is considered.

Figure 15 shows that the GAM has excellent results in all of the author's datasets, and that results are consistent across the datasets. While the Tulsa model has a problem fitting some of the authors' results, it can be noted that both the GAM and the GBM are competitive when both are calibrated correctly. Nevertheless, as stated in this document, the GAM offers a greater degree of tools such as p-values and confidence intervals as a risk measure and graphical splines to corroborate the predictors' effect. Moreover, although the mechanistic model offers much more information about fluid dynamics, it cannot predict the pressure gradient with the same level of accuracy as at both Machine Learning models.

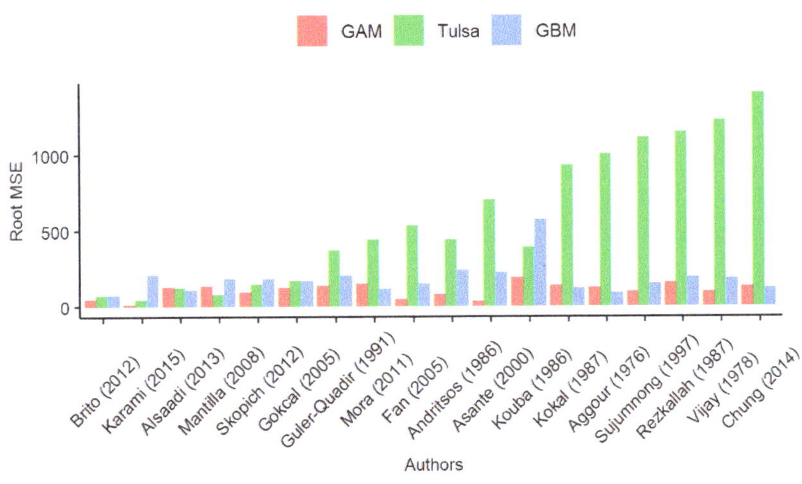

Figure 15. Model comparison using root mean square error for each author in the database. For the ML models, an out-of-sample test is used.

6. Conclusions and Future Work

This study proposed a Generalized Additive Model based on 21 recognized dimensionless numbers to predict the pressure drop of gas–liquid two-phase flow based on 5011 records in the Tulsa Unified Fluid Flow Project (TUFFP) database. The prediction results were satisfactory, obtaining a mean relative error of at most 19.98% for stratified flow and 12.93% for all the data points in a randomly sampled test set. The adjusted R^2 was 99.1%, which is the same as the deviance explained in the model. The GAM can also predict bubbly flow with a 4.3% relative error without even being trained to know the flow pattern, assuming the cases are mixed with intermittent flow. A regularization methodology was implemented to reduce the dimensionality, but the number of features did not change significantly. The dimensionless numbers used are probably the simplest ones since they use superficial velocities and nominal diameters.

This approach could be improved considering hybrid model approaches to calculate geometrical considerations such as hydraulic diameters, wetted perimeter, or mean velocities. Using these quantities in the input dimensionless numbers of the GAM should condense the information better and reduce the number of features. Finally, it is restated that GAM can also be used to improve the mechanistic model approach. Replacing closure laws with GAM-based methodologies could give a new dynamism and ideas to propose phenomenological reasons for data behavior.

7. Supporting Information

Based on the model proposed in this paper, we developed an app known as DropTool that is available at the following link: https://droptool.shinyapps.io/Pressure_Drop_App/ (accessed on January 2021). This app describes the GAM model's central core regarding the experimental data, dimensionless numbers, and method implemented. This app allows the user to estimate the pressure drop based on essential fluid and operation properties for the liquid-gas phase flow. As mentioned above, the model requires the flow pattern, and the user needs to choose from Annular, Bubble, Intermittent, or Segregated flow patterns. DropTool incorporates the tool ProPatt: *Probabilistic approach of a flow pattern map for horizontal, vertical, and inclined pipelines*, which is based on the work of Amaya-Gómez et al. [11].

Author Contributions: Conceptualization, A.C.-V., R.A.-G., M.A., C.T., C.V. and N.R.; methodology, A.C.-V., R.A.-G., C.V. and N.R.; software, A.C.-V.; validation, A.C.-V.; formal analysis, A.C.-V.; resources, C.V. and N.R.; data curation, A.C.-V. and C.T.; writing—original draft preparation, A.C.-V. and R.A.-G.; writing—review and editing, A.C.-V., R.A.-G., C.V. and N.R.; visualization, A.C.-V.; supervision, R.A.-G., M.A., C.T., C.V. and N.R. All authors have read and agreed to the published version of the manuscript.

Funding: This research received no external funding.

Institutional Review Board Statement: Not applicable.

Informed Consent Statement: Not applicable.

Data Availability Statement: Not applicable.

Conflicts of Interest: The authors declare no conflict of interest.

Nomenclature

β	linear regression coefficient
χ	Lockhart Martinelli Parameter
$\Delta\rho$	liquid density minus gas density [kg/m^3]
ΔP	Pressure Gradient [Pa/m]
$\dot{m}$	mass flow rate [kg/s]
ϵ	relative error
μ	Viscosity [Pa · s]
ρ	density [kg/m^3]
σ	superficial tension [N/m]
θ	pipe inclination [deg]
A	pipe cross-sectional area [m^2]
D	pipe internal diameter [m]
E_o	Eötvös Number
f	Friction factor
F_t	Froude Rate
Fr	Froude Number
g	gravitational acceleration [m/s^2]
L	pipe length [m]
N_μ	Viscous number
P	Pressure [Pa]
Re	Reynolds Number

T	Temperature [°C]
V	Velocity [m/s]
We	Weber Number
Y	response variable, cube root pressure gradient $[(Pa/m)^{1/3}]$
Y_{TD}	Taitel Dukler dimensionless number
AIC	Akaike information criterion
E.dof	Estimated degrees of freedom of each spline
R.dof	Reference degrees of freedom for each spline

References

1. Li, X.; Miskimins, J.L.; Hoffman, B.T. A Combined Bottom-hole Pressure Calculation Procedure Using Multiphase Correlations and Artificial Neural Network Models. In Proceedings of the SPE Annual Technical Conference and Exhibition, Amsterdam, The Netherlands, 27–29 October 2014. [CrossRef]
2. Hagedorn, A.; Brown, K. Experimental Study of Pressure Gradients Occurring During Continuous Two-Phase Flow in Small-Diameter Vertical Conduits. *J. Pet. Technol.* **1965**, *17*, 475–484. [CrossRef]
3. Beggs, D.; Brill, J. A Study of Two-Phase Flow in Inclined Pipes. *J. Pet. Technol.* **1973**, *25*, 607–617. [CrossRef]
4. Xiao, J.J.; Shonham, O.; Brill, J.P. A Comprehensive Mechanistic Model for Two-Phase Flow in Pipelines. In Proceedings of the SPE Annual Technical Conference and Exhibition, New Orleans, LA, USA, 23–26 September 1990. [CrossRef]
5. Kanin, E.; Osiptsov, A.; Vainshtein, A.; Burnaev, E. A predictive model for steady-state multiphase pipe flow: Machine learning on lab data. *J. Pet. Sci. Eng.* **2019**, *180*, 727–746. [CrossRef]
6. Carrizales, M.J.; Jaramillo, J.E.; Fuentes, D. Prediction of Multiphase Flow in Pipelines: Literature Review. *Ing. Cienc.* **2015**, *11*, 213–233. [CrossRef]
7. Ishii, M.; Hibiki, T. Drift Flux Model. In *Thermo-Fluid Dynamics of Two-Phase Flow*; Springer: New York, NY, USA, 2011; Chapter 13, pp. 361–395. [CrossRef]
8. Shi, H.; Holmes, J.A.; Durlofsky, L.J.; Aziz, K.; Diaz, L.R.; Alkaya, B.; Oddie, G. Drift-Flux Modeling of Two-Phase Flow in Wellbores. *SPE J.* **2005**, *10*, 24–33. [CrossRef]
9. Taitel, Y.; Dukler, A.E. A model for predicting flow regime transitions in horizontal and near horizontal gas-liquid flow. *AIChE J.* **1976**, *22*, 47–55. [CrossRef]
10. Zhang, H.Q.; Wang, Q.; Sarica, C.; Brill, J. Unified Model for Gas-Liquid Pipe Flow via Slug Dynamics—Part 1: Model Development. *J. Energy Resour. Technol.* **2003**, *125*, 266–273. [CrossRef]
11. Amaya-Gómez, R.; López, J.; Pineda, H.; Urbano-Caguasango, D.; Pinilla, J.; Ratkovich, N.; Muñoz, F. Probabilistic approach of a flow pattern map for horizontal, vertical, and inclined pipes. *Oil Gas Sci. Technol. Rev. IFP Energies Nouv.* **2019**, *74*, 67. [CrossRef]
12. Ali, N.; Viggiano, B.; Tutkun, M.; Bayoán Cal, R. Cluster-based reduced-order descriptions of two phase flows. *Chem. Eng. Sci.* **2020**, *222*, 115660. [CrossRef]
13. Xie, T.; Ghiaasiaan, S.; Karrila, S. Artificial neural network approach for flow regime classification in gas–liquid–fiber flows based on frequency domain analysis of pressure signals. *Chem. Eng. Sci.* **2004**, *59*, 2241–2251. [CrossRef]
14. Cozin, C.; Vicencio, F.E.C.; de Almeida Barbuto, F.; Morales, R.M.; da Silva, M.; Arruda, L. Two-Phase Slug Flow Characterization Using Artificial Neural Networks. *IEEE Trans. Instrum. Meas.* **2016**, *65*, 494–501. [CrossRef]
15. Gao, Z.K.; Jin, N.D. Characterization of chaotic dynamic behavior in the gas–liquid slug flow using directed weighted complex network analysis. *Phys. A Stat. Mech. Its Appl.* **2012**, *391*, 3005–3016. [CrossRef]
16. Ansari, A.; Sylvester, N.; Sarica, C.; Shoham, O.; Brill, J. A Comprehensive Mechanistic Model for Upward Two-Phase Flow in Wellbores. *SPE Prod. Facil.* **1994**, *9*, 143–151. [CrossRef]
17. Zhang, H.Q.; Wang, Q.; Sarica, C.; Brill, J.P. Unified Model for Gas-Liquid Pipe Flow via Slug Dynamics—Part 2: Model Validation. *J. Energy Resour. Technol.* **2003**, *125*, 274–283. [CrossRef]
18. Hernández, J.; Valencia, C.; Ratkovich, N.; Torres, C.; Muñoz, F. Data driven methodology for model selection in flow pattern prediction. *Heliyon* **2019**, *5*, e02718. [CrossRef]
19. Mohammadi, S.; Papa, M.; Pereyra, E.; Sarica, C. Genetic algorithm to select a set of closure relationships in multiphase flow models. *J. Pet. Sci. Eng.* **2019**, *181*, 106224. [CrossRef]
20. James, G.; Witten, D.; Hastie, T.; Tibshirani, R. *An Introduction to Statistical Learning: With Applications in R*; Springer Texts in Statistics; Springer: New York, NY, USA, 2013.
21. Hastie, T.; Tibshirani, R. Generalized Additive Models: Some Applications. *J. Am. Stat. Assoc.* **1987**, *82*, 371–386. [CrossRef]
22. Petalas, N.; Aziz, K. A Mechanistic Model for Multiphase Flow in Pipes. *J. Can. Pet. Technol.* **2000**, *39*. [CrossRef]
23. Ozbayoglu, E.; Ozbayoglu, M. Estimating Flow Patterns and Frictional Pressure Losses of Two-Phase Fluids in Horizontal Wellbores Using Artificial Neural Networks. *Pet. Sci. Technol.* **2009**, *27*, 135–149. [CrossRef]
24. Wood, S. *Generalized Additive Models: An Introduction with R*; Chapman and Hall/CRC: Boca Raton, FL, USA, 2006.
25. Alsaadi, Y. Liquid Loading in Highly Deviated Gas Wells. Master's Thesis, The University of Tulsa, Tulsa, OK, USA, 2013.
26. Wood, S. Thin plate regression splines. *J. R. Stat. Soc. Ser. B (Stat. Methodol.)* **2003**, *65*, 95–114. [CrossRef]
27. Aggour, M. Hydrodynamics and Heat Transfer in Two-Phase Two-Component Flows. Ph.D. Thesis, University of Manitoba, Winnipeg, MB, Canada, 1978.

28. Brito, R. Effect of Medium Oil Viscosity on Two-Phase Oil-Gas Flow Behavior in Horizonal Pipes. Master's Thesis, The University of Tulsa, Tulsa, OK, USA, 2012.
29. Fan, Y. An Investigation of Low Liquid Loading Gas-Liquid Stratified Flow in Near-Horizontal Pipes. Ph.D. Thesis, The University of Tulsa, Tulsa, OK, USA, 2005.
30. Mantilla, I. Mechanistic Modeling of Liquid Entrainment in Gas in Horizontal Pipes. Ph.D. Thesis, The University of Tulsa, Tulsa, OK, USA, 2008.
31. Andritsos, N. Effect of Pipe Diameter and Liquid Viscosity on Horizontal Stratified Flow. Ph.D. Thesis, University of Illinois at Urbana-Champaign, Urbana, IL, USA, 1986.
32. Asante, B. Multiphase Transport of Gas and Low Loads of Liquids in Pipelines. Ph.D. Thesis, University of Calgary, Calgary, AB, Canada, 2000.
33. Chung, S.; Pereyra, E.; Sarica, C.; Soto, G.; Alruhaimani, F.; Kang, J. Effect of high oil viscosity on oil-gas flow behavior in vertical downward pipes. In Proceedings of the 10th North American Conference on Multiphase Production Technology, Banff, AB, Canada, 8–10 June 2016; BHR-2016-259.
34. Gokcal, B. An Experimental and Theoretical Investigation of Slug Flow for High Oil Viscosity in Horizontal Pipes. Ph.D. Thesis, The University of Tulsa, Tulsa, OK, USA, 2008.
35. Güler-Quadir, N. Two-Phase Pressure Drop and Holdup in Flows Through Large Diameter Vertical Tubing. Ph.D. Thesis, The University of Tulsa, Tulsa, OK, USA, 1991.
36. Karami Mirazizi, H. Low Liquid Loading Three-Phase Flow and Effects of MEG on Flow Behavior. Ph.D. Thesis, The University of Tulsa, Tulsa, OK, USA, 2015.
37. Kokal, S.L. An Experimental Study of Two Phase Flow in Inclined Pipes. Ph.D. Thesis, University of Calgary, Calgary, AB, Canada, 1987. [CrossRef]
38. Kouba, G. Horizontal Slug Flow Modelling and Metering. Ph.D. Thesis, The University of Tulsa, Tulsa, OK, USA, 1986.
39. Rezkallah, K. Heat Transfer and Hydrodynamics in Two-Phase Two-Component Flow in a Vertical Tube. Ph.D. Thesis, University of Manitoba, Winnipeg, MB, Canada, 1987.
40. Skopich, A. Experimental Study of Surfactant Effect on Liquid Loading in 2-in and 4-in Diameter Vertical Pipes. Master's Thesis, The University of Tulsa, Tulsa, OK, USA, 2012.
41. Sujumnong, M. Heat Transfer Pressure Drop and Void Fraction in Two Phase, Two-Component Flow in a Vertical Tube. Ph.D. Thesis, University of Manitoba, Winnipeg, MB, Canada, 1997.
42. Vijay, M. A Study of Heat Transfer in Two-Phase Two-Component Flow in Vertical Tube. Ph.D. Thesis, University of Manitoba, Winnipeg, MB, Canada, 1977.
43. Graham, D.; Kopke, H.; Wilson, M.; Yashar, D. *An Investigation of Void Fraction in the Stratified/Annular Flow Regions in Smooth, Horizontal Tubes*; Technical Report; Part of ACRC Project 74; Air Conditioning and Refrigeration Center, College of Engineering, University of Illinois at Urbana-Champaign: Urbana, IL, USA, 1999.
44. Brauner, N.; Maron, D. Identification of the range of 'small diameters' conduits, regarding two-phase flow pattern transitions. *Int. Commun. Heat Mass Transf.* **1992**, *19*, 29–39. [CrossRef]
45. Al-Safran, E.; Kora, C.; Sarica, C. Prediction of slug liquid holdup in high viscosity liquid and gas two-phase flow in horizontal pipes. *J. Pet. Sci. Eng.* **2015**, *133*, 566–575. [CrossRef]
46. Marra, G.; Wood, S. Practical variable selection for generalized additive models. *Comput. Stat. Data Anal.* **2011**, *55*, 2372–2387. [CrossRef]
47. Andritsos, N.; Williams, L.; Hanratty, T. Effect of liquid viscosity on the stratified-slug transition in horizontal pipe flow. *Int. J. Multiph. Flow* **1989**, *15*, 877–892. [CrossRef]

processes

MDPI

Article

Stability of a Viscous Liquid Jet in a Coaxial Twisting Compressible Airflow

Li-Mei Guo, Ming Lü * and Zhi Ning

School of Mechanical, Electronic and Control Engineering, Beijing Jiaotong University, Beijing 100044, China; 18121390@bjtu.edu.cn (L.-M.G.); zhining@bjtu.edu.cn (Z.N.)
* Correspondence: lvming@bjtu.edu.cn

Abstract: Based on the linear stability analysis, a mathematical model for the stability of a viscous liquid jet in a coaxial twisting compressible airflow has been developed. It takes into account the twist and compressibility of the surrounding airflow, the viscosity of the liquid jet, and the cavitation bubbles within the liquid jet. Then, the effects of aerodynamics caused by the gas–liquid velocity difference on the jet stability are analyzed. The results show that under the airflow ejecting effect, the jet instability decreases first and then increases with the increase of the airflow axial velocity. When the gas–liquid velocity ratio $A = 1$, the jet is the most stable. When the gas–liquid velocity ratio $A > 2$, this is meaningful for the jet breakup compared with $A = 0$ (no air axial velocity). When the surrounding airflow swirls, the airflow rotation strength E will change the jet dominant mode. E has a stabilizing effect on the liquid jet under the axisymmetric mode, while E is conducive to jet instability under the asymmetry mode. The maximum disturbance growth rate of the liquid jet also decreases first and then increases with the increase of E. The liquid jet is the most stable when $E = 0.65$, and the jet starts to become more easier to breakup when $E = 0.8425$ compared with $E = 0$ (no swirling air). When the surrounding airflow twists (air moves in both axial and circumferential directions), given the axial velocity to change the circumferential velocity of the surrounding airflow, it is not conducive to the jet breakup, regardless of the axisymmetric disturbance or asymmetry disturbance.

Keywords: jet stability; swirl; twist; dominant mode

Citation: Guo, L.-M.; Lü, M.; Ning, Z. Stability of a Viscous Liquid Jet in a Coaxial Twisting Compressible Airflow. *Processes* **2021**, 9, 918. https://doi.org/10.3390/pr9060918

Academic Editor: Jean-Claude Assaf

Received: 19 April 2021
Accepted: 19 May 2021
Published: 24 May 2021

1. Introduction

A liquid jet is a common phenomenon in various fields, including rainfall in nature, pesticide spraying in agriculture, and fuel injection of internal combustion engines in engineering, and so on. Therefore, the instability mechanism of the liquid jet has an important academic significance and engineering value.

At present, many studies have confirmed that the liquid jet in a coaxial high-speed airflow can achieve efficient mixing of the liquid jet and the surrounding gas. Compared with the liquid jet in a stationary gas, the liquid jet in the high-speed airflow is more conducive to atomization, and some research works have been obtained about the stability of a liquid jet in a coaxial airflow [1–8]. Sometimes, in the process of spraying and atomization, it is often accompanied by the swirling of the surrounding airflow [9,10]. Actually, the swirling movement of the surrounding airflow has a complex effect on the jet instability, which has gradually become a research hotspot [11–20]. Jog and Ibrahim [11] studied the swirling effect with the neglect of airflow compressibility. Lin et al. [12] studied the swirling effect with the neglect of the liquid viscosity and the airflow compressibility. Du et al. [14,15] studied the swirling effect with the neglect of the twisting motion and compressibility of the surrounding airflow. Strasser and Battaglia [16] studied the swirling effect and the compressibility of the jet based on the large-eddy simulation. Lü et al. [17,18] studied the swirling and compressibility effect of the surrounding airflow with the neglect of the liquid viscosity and the twisting motion of the surrounding airflow.

In addition, the effect of liquid viscosity on jet stability cannot be ignored [20–22]. Chandrasekhar [23] and Decent [24] studied the effect of fluid viscosity on the stability of liquid jet, and found that the fluid viscosity played an important role in promoting jet instability. Lin [25] mainly considered the viscosity of the liquid jet and analyzed the jet stability under different liquid viscosities. Yan [26,27] considered both the fluid viscosity and the airflow compressibility in order to analyze the jet stability. However, both of them neglected the swirling of the surrounding airflow. In addition, in the process of liquid injection, such as fuel injection in the internal combustion engines, cavitation bubbles always exist in the fuel jet leaving the nozzle [17,18]. Cavitation bubbles turn the liquid jet into a gas–liquid two-phase flow, which increases the instability of the liquid jet. Therefore, the study of the effect of cavitation bubbles on jet instability is of significance.

However, because of the complexity of all of these problems, scholars often neglect the coupling effect of the twist (movement both in axial and circumferential direction) and compressibility of the surrounding airflow, the viscosity of the liquid jet, and the cavitation bubbles within the liquid jet, which makes the research results deviate from the actual conditions.

According to current research situations and existing problems, on the premise of a comprehensive consideration of the twist and compressibility of the surrounding airflow, the viscosity of the liquid jet, and the cavitation bubbles within the liquid jet, the dispersion equation for the stability of a viscous liquid jet in a coaxial twisting compressible airflow has been developed. On this basis, the effects of airflow ejection, airflow swirling, and airflow twist on the stability of the liquid jet are discussed.

2. Mathematical Model

2.1. Physical Model and Initial Flow Field

A bundle of viscous liquid is injected into a coaxial twisting compressible gas medium from a cylindrical nozzle with a radius of a. The cylindrical coordinate system is established at the outlet of the nozzle, and the jet direction is opposite to the z-axis direction. The liquid jet is assumed to have a radius a and initial velocity U_0, while the surrounding gas has velocity in the z-axis direction U_2 and airflow rotation strength W_0, as shown in Figure 1.

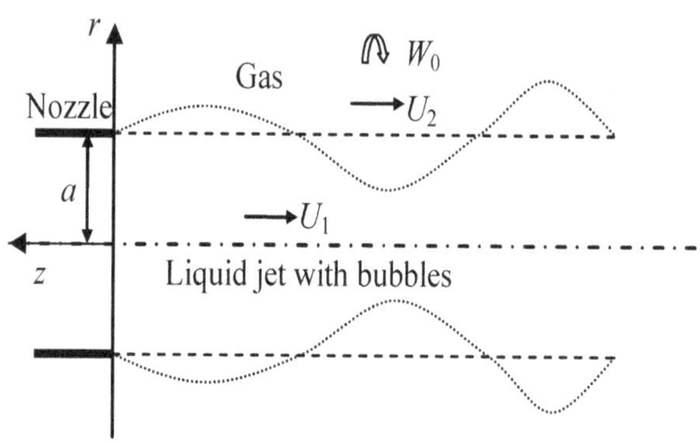

Figure 1. Physical model.

Based on the above physical model, the following assumptions are made:

(1) The surrounding gas is a compressible Newtonian fluid;
(2) Ignore the influence of the viscosity, gravity, and temperature of the surrounding gas;
(3) The liquid jet does not swirl, and the surrounding gas has a coaxial twisting velocity;
(4) There is no slippage between the cavitation bubbles and the liquid jet, and there is no interaction between the cavitation bubbles;
(5) The mixed phase consisting of uniformly distributed cavitation bubbles and liquid jet is a continuous medium.

In the coordinate system shown in Figure 1, the basic flow field (liquid jet velocity, the surrounding airflow velocity, the pressure difference between the liquid jet and the surrounding airflow, jet density, jet viscosity, and sound velocity in the liquid jet) is established based on the above assumptions, as follows:

$$\overline{\mathbf{V}}_1 = (0,\, 0, -U_1) \qquad (0 \le r \le a) \tag{1}$$

$$\overline{\mathbf{V}}_2 = (0,\, W_0/r, -U_2) \qquad (a < r < \infty) \tag{2}$$

$$\overline{p}_2(r) - \overline{p}_1 = p_\sigma + p_{int} = -\frac{\sigma}{a} + \frac{1}{2}\overline{\rho}_2 W_0^2\left(\frac{1}{a^2} - \frac{1}{r^2}\right) \quad (a < r < \infty) \tag{3}$$

$$\overline{\rho}_1 = \alpha\rho_v + (1-\alpha)\rho_l \tag{4}$$

$$\mu_1 = \alpha\mu_v + (1-\alpha)\mu_l \tag{5}$$

where $\overline{p}_2(r)$ represent the surrounding gas pressure; $\overline{p}_1$ represent the liquid jet pressure; p_σ is the surface tension; p_{in} is the gas phase inertia; σ is the surface tension factor at the interface; $\overline{\rho}_2$ is the gas density; $\overline{\rho}_1$ is the liquid jet density; W_0 is the gas rotation strength; ρ_v and ρ_l are cavitation bubble density and liquid density, respectively; μ_v and μ_l are the cavitation bubble viscosity and liquid viscosity, respectively; α is the bubble volume fraction, $\alpha = 4\pi r_v^3 N/3$; r_v is the cavity average radius; and N is the number of cavitations per unit volume.

2.2. Establishment and Solution of Mathematical Model

The equation is based on the linear stability analysis method, which is based on the establishment of the airflow and liquid fluid disturbance governing equations and the determination of the boundary conditions, using the linear small disturbance method to ignore the high-order small quantities to linearize the equation, build and solve the equations, and then carry out research.

(1) Disturbance governing equations of compressible twisting airflows

Consider the compressibility and neglect the viscosity and gravity of the surrounding airflow, the airflow satisfies the following continuity equation and Euler equation [28]:

$$\frac{\mathrm{D}\rho_2}{\mathrm{D}t} + \rho_2\left(\frac{1}{r}\frac{\partial}{\partial r}(rv_{r2}) + \frac{1}{r}\frac{\partial v_{\theta 2}}{\partial\theta} + \frac{\partial v_{z2}}{\partial z}\right) = 0 \tag{6}$$

$$\begin{cases} \frac{\mathrm{D}v_{r2}}{\mathrm{D}t} - \frac{v_{\theta 2}^2}{r} = -\frac{1}{\rho_2}\frac{\partial p_2}{\partial r} \\ \frac{\mathrm{D}v_{\theta 2}}{\mathrm{D}t} + \frac{v_{r2}v_{\theta 2}}{r} = -\frac{1}{\rho_2 r}\frac{\partial p_2}{\partial\theta} \\ \frac{\mathrm{D}v_{z2}}{\mathrm{D}t} = -\frac{1}{\rho_2}\frac{\partial p_2}{\partial z} \end{cases} \tag{7}$$

where subscript 2 represents airflow parameters, and v_r, v_θ, and v_z are the radial, circumferential, and axial velocities, respectively.

Perturbation analysis and linearization of Equations (6) and (7), we can get the surrounding airflow disturbance governing equations as the following form:

$$\frac{\partial(p'_2/c_2^2)}{\partial t} + \frac{W_0}{r^2}\frac{\partial(p'_2/c_2^2)}{\partial\theta} - U_2\frac{\partial(p'_2/c_2^2)}{\partial z} + \overline{\rho}_2\left(\frac{v'_{r2}}{r} + \frac{\partial v'_{r2}}{\partial r} + \frac{1}{r}\frac{\partial v'_{\theta 2}}{\partial\theta} + \frac{\partial v'_{z2}}{\partial z}\right) = 0 \tag{8}$$

$$\begin{cases} \overline{\rho}_2\left(\frac{\partial v'_{r2}}{\partial t} + \frac{W_0}{r^2}\frac{\partial v'_{r2}}{\partial\theta} - U_2\frac{\partial v'_{r2}}{\partial z} - \frac{2W_0 v'_{\theta 2}}{r^2}\right) = -\frac{\partial p'_2}{\partial r} \\ \overline{\rho}_2\left(\frac{\partial v'_{\theta 2}}{\partial t} + \frac{W_0}{r^2}\frac{\partial v'_{\theta 2}}{\partial\theta} - U_2\frac{\partial v'_{\theta 2}}{\partial z}\right) = -\frac{1}{r}\frac{\partial p'_2}{\partial\theta} \\ \overline{\rho}_2\left(\frac{\partial v'_{z2}}{\partial t} + \frac{W_0}{r^2}\frac{\partial v'_{z2}}{\partial\theta} - U_2\frac{\partial v'_{z2}}{\partial z}\right) = -\frac{\partial p'_2}{\partial z} \end{cases} \tag{9}$$

where the apostrophe indicates a small disturbance parameter.

Equations (8) and (9) include four equations, but contain five unknown parameters. In order to close the four equations, use $\partial p'_2/\partial\rho'_2 = c_2^2$ to relate the disturbance density and pressure with the speed of sound. At this point, the airflow disturbance governing equations are established.

(2) Disturbance governing equations of viscous liquid jets

Consider the viscosity and neglect the compressibility and gravity of the liquid jet, the liquid jet satisfies the following continuity equation and momentum equation [28]:

$$\frac{1}{r}\frac{\partial}{\partial r}(rv_{r1})+\frac{1}{r}\frac{\partial v_{\theta 1}}{\partial \theta}+\frac{\partial v_{z1}}{\partial z}=0 \tag{10}$$

$$\begin{cases} \frac{Dv_{r1}}{Dt}-\frac{v_{\theta 1}^2}{r}=-\frac{1}{\rho_1}\frac{\partial p_1}{\partial r}+\nu_1\left(\nabla^2 v_{r1}-\frac{v_{r1}}{r^2}-\frac{2}{r^2}\frac{\partial v_{\theta 1}}{\partial \theta}\right) \\ \frac{Dv_{\theta 1}}{Dt}+\frac{v_{r1}v_{\theta 1}}{r}=-\frac{1}{\rho_1 r}\frac{\partial p_1}{\partial \theta}+\nu_1\left(\nabla^2 v_{\theta 1}+\frac{2}{r^2}\frac{\partial v_{r1}}{\partial \theta}-\frac{v_{\theta 1}}{r^2}\right) \\ \frac{Dv_{z1}}{Dt}=-\frac{1}{\rho_1}\frac{\partial p_1}{\partial z}+\nu_1\nabla^2 v_{z1} \end{cases} \tag{11}$$

where subscript 1 represents the liquid jet parameters and ν_1 is the liquid jet kinematic viscosity.

From the perturbation analysis and linearization of Equations (10) and (11), we can obtain the following disturbance governing equations of the liquid jet:

$$\frac{v'_{r1}}{r}+\frac{\partial v'_{r1}}{\partial r}+\frac{1}{r}\frac{\partial v'_{\theta 1}}{\partial \theta}+\frac{\partial v'_{z1}}{\partial z}=0 \tag{12}$$

$$\frac{\partial v'_{r1}}{\partial t}-U_1\frac{\partial v'_{r1}}{\partial z}=-\frac{1}{\overline{\rho}_1}\frac{\partial p'_1}{\partial r}+\nu_1\left(\nabla^2 v'_{r1}-\frac{v'_{r1}}{r^2}-\frac{2}{r^2}\frac{\partial v'_{\theta 1}}{\partial \theta}\right) \tag{13}$$

$$\frac{\partial v'_{\theta 1}}{\partial t}-U_1\frac{\partial v'_{\theta 1}}{\partial z}=-\frac{1}{\overline{\rho}_1 r}\frac{\partial p'_1}{\partial \theta}+\nu_1\left(\nabla^2 v'_{\theta 1}+\frac{2}{r^2}\frac{\partial v'_{r1}}{\partial \theta}-\frac{v'_{\theta 1}}{r^2}\right) \tag{14}$$

$$\frac{\partial v'_{z1}}{\partial t}-U_1\frac{\partial v'_{z1}}{\partial z}=-\frac{1}{\overline{\rho}_1}\frac{\partial p'_1}{\partial z}+\nu_1\nabla^2 v'_{z1} \tag{15}$$

(3) Boundary conditions

At the interface between the liquid jet and the surrounding gas, the boundary conditions include the kinematic boundary conditions and the dynamic boundary conditions:

$$v_{ri}=\frac{\partial \eta}{\partial t}+\frac{v_{\theta i}}{r}\frac{\partial \eta}{\partial \theta}+v_{zi}\frac{\partial \eta}{\partial z} \tag{16}$$

$$p_1-p_2=2\mu_1\frac{\partial v_{r1}}{\partial r}-\sigma\left(\frac{1}{r_1}+\frac{1}{r_2}+\frac{1}{a}\right)+\frac{1}{2}\overline{\rho}_2 W_0^2\left(\frac{1}{a^2}-\frac{1}{(a+\eta)^2}\right) \tag{17}$$

where η is the perturbation on the liquid jet at the interface.

Perturbation analysis and linearization of Equations (16) and (17) can obtain the following boundary conditions:

$$\overline{v}_{ri}+v'_{ri}=\frac{\partial \eta}{\partial t}+\frac{\overline{v}_{\theta i}}{a+\eta}\frac{\partial \eta}{\partial \theta}+\overline{v}_{zi}\frac{\partial \eta}{\partial z} \tag{18}$$

$$a^4\left(p'_1-p'_2\right)=2a^4\mu_1\frac{\partial v'_{r1}}{\partial r}-a^2\sigma\left(\eta+a^2\frac{\partial^2 \eta}{\partial z^2}+\frac{\partial^2 \eta}{\partial \theta^2}\right)+\overline{\rho}_2 W_0^2 a\eta \tag{19}$$

(4) Dispersion equation

Through the above disturbance governing Equations (8), (9), (12)–(15) and boundary conditions (18)–(19), a homogeneous linear algebraic equations is established:

$$AX=0 \tag{20}$$

where $X=[a_{11}, a_{12}, d_{11}, d_{22}, \eta_0]^T$, and A is a 5×5 coefficient matrix containing k, ω, m, and other jet parameters:

$$A = \begin{bmatrix} A_{11} & A_{12} & A_{13} & A_{14} & A_{15} \\ A_{21} & A_{22} & A_{23} & A_{24} & A_{25} \\ A_{31} & A_{32} & A_{33} & A_{34} & A_{35} \\ A_{41} & A_{42} & A_{43} & A_{44} & A_{45} \\ A_{51} & A_{52} & A_{53} & A_{54} & A_{55} \end{bmatrix}$$

The expressions for each element in matrix A are shown in the Appendix A.

The condition for the existence of a non-zero solution of the Equation (20) is that the determinant of the coefficient matrix is 0:

$$|A| = 0 \tag{21}$$

Equation (21) is the dispersion equation describing the stability of a viscous liquid jet in a compressible twisting airflow. In view of the complexity of the dispersion equation, this paper gives the following abbreviated form:

$$f\ (k, \omega, m, We, Re_1, E, A, Ma_2, Q) = 0 \tag{22}$$

where $k = k_r + ik_i$, k_r is the wave number in the z direction, and the relationship with the wavelength λ is $k_r = 2\pi a/\lambda$, k_i is the disturbance spatial growth rate; $\omega = \omega_r + i\omega_i$, ω_r is the disturbance temporal growth rate and ω_i is the wave frequency; m is the wave number in the θ direction; $We = \sigma/(\alpha\rho_v + (1-\alpha)\rho_l)U_1^2 a$, note that We is the reciprocal of the Weber number and signifies the ratio of the surface tension to the inertial force; $Re_1 = U_1 a/\nu_1$ is the liquid jet Reynolds number, which reflects the ratio of inertial force to viscous force; $E = W_0/(U_0 a)$ is the non-dimensional rotational strength; $A = U_2/U_1$, is the gas–liquid axial velocity ratio; $Ma_2 = U_0/c_2$ is the surrounding gas Mach number; and $Q = \overline{\rho}_2/(\alpha\rho_v + (1-\alpha)\rho_l)$ is the gas–liquid density ratio.

The effects of the liquid jet viscosity, cavitation bubbles within the liquid jet, and the twist and compressibility of the surrounding airflow are considered in the above established dispersion Equation (22). In addition, the secant method is used to solve this dispersion equation.

The mathematical modal is based on a simplified mathematical model that assumes (as valid) a modal and stability analysis. Through theoretical analysis, the mechanism of jet breakup and atomization can be explained fundamentally. However, there is a certain difference compared with the actual jet column; the next step will be studied through direct numerical simulations or experiments. Furthermore, the bubble volume fraction bubble has a certain value range. If $\alpha > 0.2$, it will not be used as a continuum, and the dispersion equation in this paper will be inapplicable.

2.3. Verification and Solution

For a special case where an inviscid liquid jet without cavitation bubbles is injected into an untwisted and incompressible airflow under axisymmetric mode, $m = 0$, $Re_1 = 0$, $E = 0$, $A = 0$, $Ma_2 = 0$, and $\alpha = 0$, then the dispersion Equation (22) is reduced to the following:

$$\frac{(\omega - ik)^2 \mathrm{I}_0(k)}{\mathrm{I}_1(k)} + \frac{Q\omega^2 \mathrm{K}_0(k)}{\mathrm{K}_1(k)} + kWe\left(k^2 - 1\right) = 0 \tag{23}$$

The reduced dispersion Equation (23) is the same as the dispersion equation derived by Lin and Lian [18].

For another special case where only the compressibility of the surrounding airflow and the cavitation bubbles within the liquid jet are neglected, that is, $Ma_2 = 0$, $\alpha = 0$, then the dispersion Equation (22) is reduced to the following:

$$a_1\left(\frac{\mathrm{I}'_m(k)}{\mathrm{I}_m(k)}\right)\left(\frac{\mathrm{I}'_m(\lambda)}{\mathrm{I}_m(\lambda)}\right)^2 + a_2\left(\frac{\mathrm{I}'_m(\lambda)}{\mathrm{I}_m(\lambda)}\right)^2 + a_3\left(\frac{\mathrm{I}'_m(k)\mathrm{I}'_m(\lambda)}{\mathrm{I}_m(k)\mathrm{I}_m(\lambda)}\right) + a_4\left(\frac{\mathrm{I}'_m(k)}{\mathrm{I}_m(k)}\right) + a_5\left(\frac{\mathrm{I}'_m(\lambda)}{\mathrm{I}_m(\lambda)}\right) + a_6 = 0 \tag{24}$$

The reduced dispersion Equation (24) is consistent with the one derived by Lin [29], and the formal difference is due to the definition of the axial direction and wave frequency.

The comparison of the above two specific cases can prove the correctness of the dispersion equation to some extent.

In order to verify the correctness of the numerical solution method, the calculating conditions and original data from [24] are compared with the present calculation results in this paper. The comparison results are shown in Figure 2.

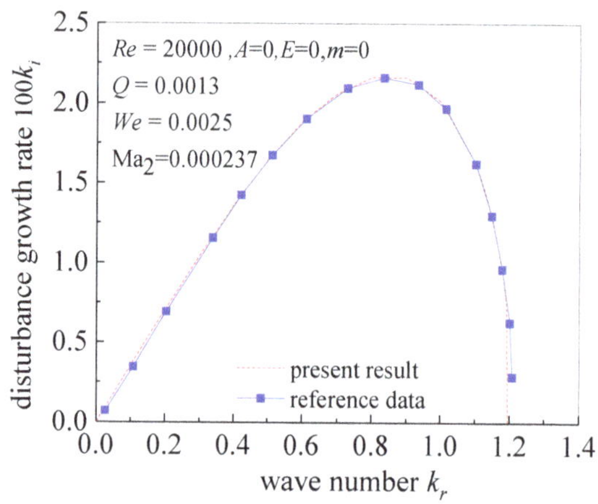

Figure 2. Comparison of the calculation results with the data in [24].

The authors of [24] provide an example that a viscous liquid jet is injected into the stationary gas under the axisymmetric disturbance. As shown in Figure 2, the calculation results in this paper are in good agreement with the data in the literature, which indicates that the numerical solution method of the dispersion equation is reasonable and effective.

3. Results and Discussions

In this paper, diesel was chosen as the liquid jet and air was chosen as the surrounding gas, and the relevant parameters used are shown in Table 1.

Table 1. Calculating parameters.

Parameters	Units	Values
Liquid density	kg/m^3	848
Surface tension	N/m	2.689×10^{-2}
Gas density	kg/m^3	1.193
Kinematic viscosity	m^2/s	7.6658×10^{-6}
Jet speed	m/s	10
Nozzle radius	m	1×10^{-4}
Temperature	K	300
Sound speed in gas	m/s	348

The dispersion Equation (22) can reflect the effects of the liquid jet viscosity, cavitation bubbles within the liquid jet, and the twist and compressibility of the surrounding airflow on jet instability. Here, we only investigate the aerodynamic effects caused by the surrounding airflow ejection (axial speed), swirl (circumferential speed), and twist (both the axial and circumferential speed) on the stability of the liquid jet when the other parameter values are given as $We = 0.0032$, $Re_1 = 130$, $Ma_2 = 0.029$, and $Q = 0.0014$.

3.1. Effect of Ejecting Airflow on the Stability of Viscous Liquid Jet

This section analyzes the effect of ejecting airflow on the stability of a viscous liquid jet. $A = U_2/U_1$ is the gas–liquid axial velocity ratio, which can be used to characterize the ejecting airflow velocity size. $A = 1$ means that the axial velocity of liquid jet is equal to the axial velocity of the surrounding airflow. According to the value of the gas–liquid axial velocity ratio A, the surrounding airflow ejection is divided into strong ejection and weak ejection. When gas–liquid axial velocity ratio $A \leq 2$, it is classified as a weak ejection. $A > 2$ is classified as the strong ejection in this paper.

Figure 3 shows that the disturbance growth rate changes with the axial wave number in the axisymmetric disturbance ($m = 0$) under the weak airflow ejection.

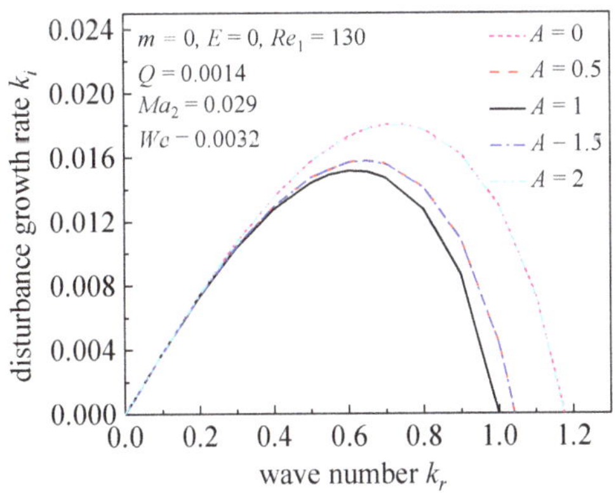

Figure 3. Variation of disturbance growth rates with axial wave numbers under the weak airflow ejection.

As shown in Figure 3, when the surrounding airflow ejection exists, and gas–liquid axial velocity ratio $A \leq 2$, the disturbance growth rate of the liquid jet decreases first and then increases with the increase of the gas–liquid velocity ratio A (from 0 to 2). When the air axial velocity is equal to the liquid axial velocity ($A = 1$), the disturbance growth rate is the smallest, and the liquid jet is the most stable. When gas–liquid axial velocity ratio $A = 2$, the disturbance growth rate of the liquid jet is basically the same as when $A = 0$ (no air axial velocity); it is also observed that the disturbance growth rate curve basically coincides when $A = 0.5$ and $A = 1.5$. So, we can get that when the absolute value of $(A - 1)$ is the same, the resulting aerodynamic effect on the liquid jet stability is basically the same. Therefore, the ejection of the surrounding airflow can suppress the breakup of the liquid jet when $A \in [0, 2]$ compared with no air axial velocity.

Figure 4 shows the comparison of the disturbance growth rates versus axial wave numbers under the strong airflow ejection ($A > 2$) in the axisymmetric disturbance mode ($m = 0$) and the asymmetry disturbance mode ($m = 1$), respectively.

It can be seen from Figure 4a that the disturbance growth rates increase sharply with the increase of the gas–liquid velocity ratios, and the range of the unstable axial wave numbers is significantly widened. This indicates that the increase in the airflow axial velocity can increase the aerodynamic effect and then promote the instability of the liquid jet when $A > 2$. From the inverse proportional relationship between the axial wave number and the droplet size, it can be concluded that the increase of the airflow axial velocity will reduce the droplet size and get the stronger atomization effect under the strong airflow ejection.

A comparison of Figure 4a,b shows that the effect of airflow ejection on the jet stability in the axisymmetric and asymmetry mode is similar. In addition, the maximum disturbance growth rate of the axisymmetric mode is always bigger than that of the asymmetry mode, but the gap between the two modes is gradually narrowing with the increasing of the gas–liquid speed ratios, which suggests that the airflow ejection may change the dominant mode of the liquid jet.

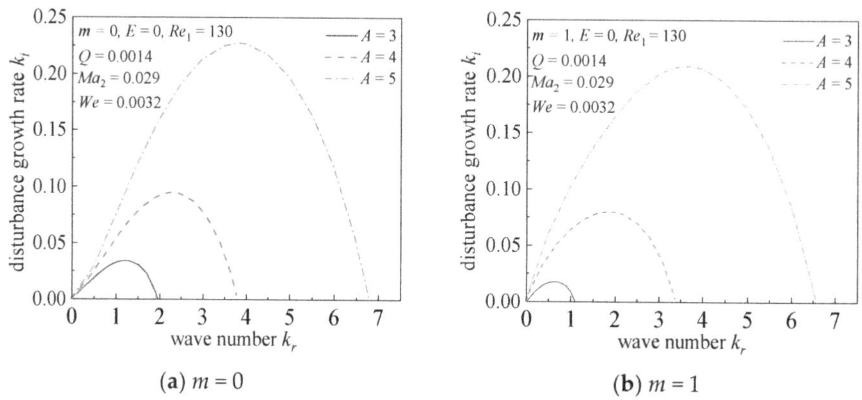

Figure 4. Comparison of the disturbance growth rates versus axial wave numbers under strong airflow ejection.

3.2. Effect of Swirling Airflow on the Stability of Viscous Liquid Jet

Some research results have shown that the swirling airflow has a complex effect on the jet stability, and it is different under the axisymmetric mode and asymmetry mode [30]. This section will analyze the influence of the aerodynamic caused by the swirl of the surrounding airflow on the jet stability. $E = W_0/(U_0 a)$ represents the non-dimensional airflow rotational strength.

Figure 5 shows the effects of the swirling airflow on jet stability under the axisymmetric mode ($m = 0$) and asymmetry mode ($m = 1$). Note that the gas–liquid velocity ratio $A = 0$.

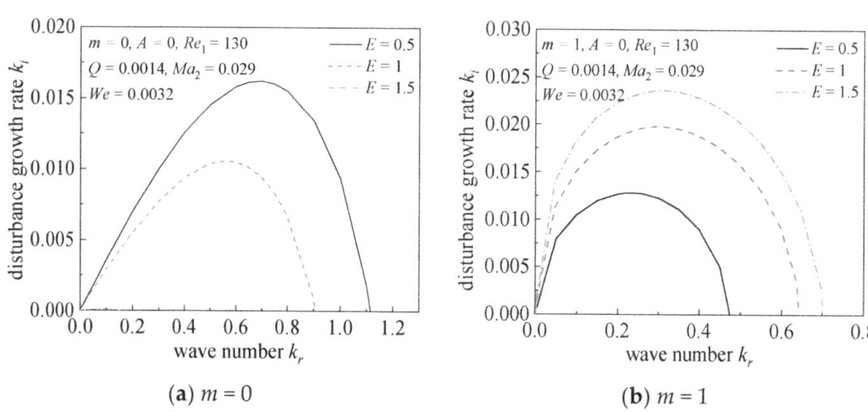

Figure 5. Effects of the swirling airflow on jet stability under the axisymmetric and asymmetry mode.

As shown in Figure 5a, under the axisymmetric mode, with the increase of the surrounding airflow rotation strength, the disturbance growth rates decrease obviously, and the range of the axial wave numbers decrease significantly. When the airflow rotation strength $E = 1.5$, the disturbance growth rate is very small, which is difficult to observe in Figure 5a. As shown in Figure 5b, under the asymmetry mode, it is found that the effect of the swirling airflow on the jet stability is completely opposite to that in Figure 5a, that is, the disturbance growth rates increases sharply with the increasing of the airflow rotation strength, which will increase the jet instability.

Comparing Figure 5a with Figure 5b, it is found that when the airflow rotation strength $E = 0.5$, the maximum disturbance growth rate of the axisymmetric disturbance $k_{i\_max} = 0.0162$, while the maximum disturbance growth rate of asymmetry disturbance $k_{i\_max} = 0.0128$, so the dominant mode is the axisymmetric mode. When the airflow rotation strength $E = 1$, the maximum disturbance growth rate of the axisymmetric disturbance $k_{i\_max} = 0.0106$, while the maximum disturbance growth rate of the asymmetry disturbance $k_{i\_max} = 0.0198$, so the asymmetry mode is dominant. It is indicated that the coaxial swirl of the surrounding airflow can change the dominant mode of the liquid jet stability, and as the rotation strength E increases, the dominant mode changes from axisymmetric mode to asymmetric mode. Table 2 shows the dominant modes under different airflow rotation strengths.

Table 2. The dominant modes under different airflow rotation strengths.

E	0	0.5	1	3	5	7
m	0	0	1	2	6	10

In order to further analyze the effects of the swirling airflow on jet stability under the dominant modes, Figure 6 gives the maximum disturbance growth rate under the dominant mode versus the different airflow rotation strength.

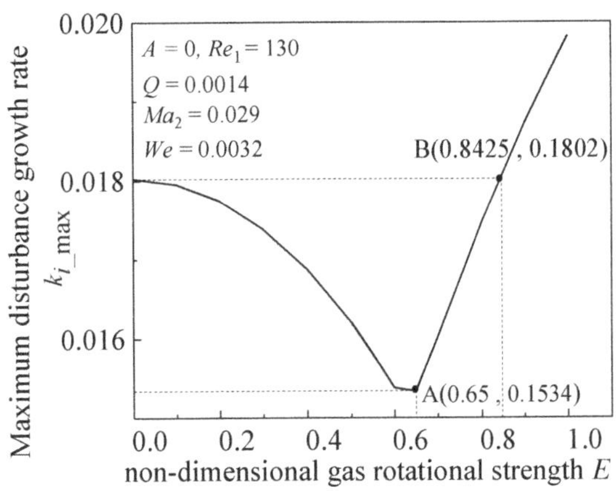

Figure 6. The maximum disturbance growth rate under the dominant mode versus the different airflow rotation strength.

It can be seen from Figure 6 that the maximum disturbance growth rate is not proportional to the airflow rotation strength. Similar to the law of airflow ejection, as the airflow rotation strength increases, the maximum disturbance growth rate decreases first and then increases sharply. The maximum disturbance growth rate is the smallest when the airflow rotation strength is at point A ($E = 0.65$, $k_{i\_max} = 0.1534$). At this point, the aerodynamic force is the smallest, and the liquid jet is the most stable. When point A is at its lowest, the coaxial swirl of the surrounding airflow can change the dominant mode of the liquid jet stability. Under axisymmetric disturbances, rotation strength is conducive to the stability of the jet, while under asymmetry disturbances, it is conducive to the instability and splitting of the jet. When the airflow rotation strength is at point B ($E = 0.8425$, $k_{i\_max} = 0.1802$), the maximum disturbance growth rate just exceeds the case of no airflow rotation ($E = 0$). According to the calculation conditions of this paper, when the surrounding airflow swirls, the increase of the airflow rotation strength when $E > 0.8425$ is beneficial to the breakup of the liquid jet.

3.3. Effect of Twisting Airflow on the Stability of Viscous Liquid Jet

To our knowledge, few scholars have studied the influence of the coupling effect on the jet stability when the surrounding airflow has both an axial and circumferential velocity. This section will analyze the influence of the aerodynamic effect caused by the twist of the surrounding airflow on the stability of the liquid jet.

Figure 7 shows the comparison of the effects of the twisting airflow on jet stability under the axisymmetric mode ($m = 0$) and asymmetry mode ($m = 1, 2, 3$). Note that the surrounding airflow has an axial velocity ($A = 6$) and has different rotation strengths ($E = 0, 0.5, 1$, and 1.5).

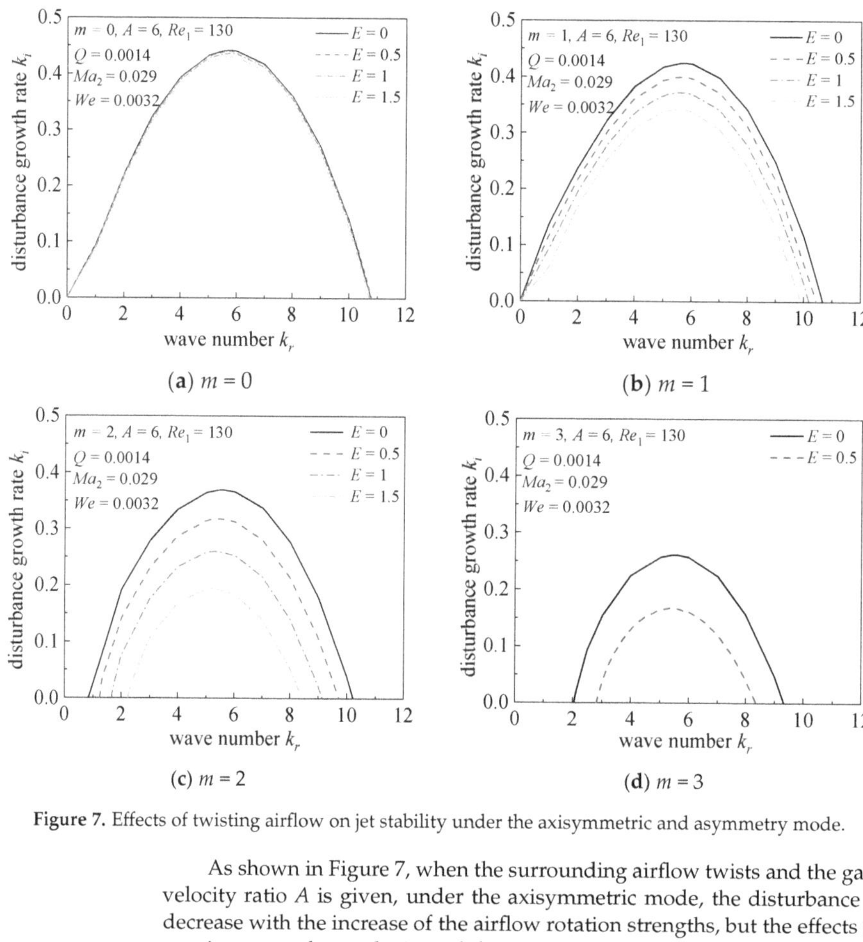

Figure 7. Effects of twisting airflow on jet stability under the axisymmetric and asymmetry mode.

As shown in Figure 7, when the surrounding airflow twists and the gas–liquid axial velocity ratio A is given, under the axisymmetric mode, the disturbance growth rates decrease with the increase of the airflow rotation strengths, but the effects of the airflow rotation strengths on the jet stability are relatively small. Under the asymmetry mode, it is different from the case that $m = 0$. When the airflow rotation strengths increase, the disturbance growth rates decrease obviously. Therefore, when the surrounding airflow twists, the stability of the liquid jet is enhanced with the increase of the airflow rotation strength both in the axisymmetric mode and in the asymmetry mode. According to the calculation conditions of this paper, the twist of the surrounding airflow is not conducive to the breakup of the liquid jet.

In addition, a comparison of Figures 5 and 7 shows that the effects of the twisting airflow on jet stability are different compared with the effects of swirling airflow on jet stability, which indicates that the gas–liquid axial velocity ratio A plays an important role in jet stability.

4. Conclusions

(1) Take into account the twist and compressibility of the surrounding airflow, the viscosity of the liquid jet and the cavitation bubbles within the liquid jet, a mathematical model for describing the stability of the liquid jet in a coaxial twisting compressible gas flow was established, the model and its solving method were verified.

(2) The ejection of surrounding airflow is divided into weak ejection ($A \leq 2$) and strong ejection ($A > 2$). Under the weak ejection of the surrounding airflow, the maximum perturbation growth rate of the jet decreases first and then increases with the increase of the axial velocity of the airflow. When the gas–liquid axial velocity is the same ($A = 1$), the maximum perturbation growth rate of the liquid jet is the smallest. When the absolute value of $(A - 1)$ is the same, the resulting aerodynamic effect on the liquid jet stability is basically the same. Under the strong ejection of the surrounding airflow, the maximum perturbation growth rate of the liquid jet increases sharply with the increase in the air ejection velocity. It is concluded that the greater the gas–liquid velocity, the more unstable the liquid jet becomes.

(3) When the surrounding airflow swirls, the swirling airflow plays a stable role on the liquid jet in the axisymmetric mode, as opposed to the asymmetry mode. The dominant mode changes from axisymmetric to asymmetry as the airflow rotation strength increases. For a cylindrical liquid jet, when the airflow rotation strength is increased, the maximum growth rate of the liquid jet decreases first and then increases. The liquid jet is the most stable when $E = 0.65$, and the liquid jet begins to become easier to breakup when $E = 0.8425$ compared with $E = 0$ (no swirling air). The increase in the airflow rotation strength from 0.8425 is beneficial to the breakup of the liquid jet, and the airflow rotation strength has a significant effect on the stability of the liquid jet.

(4) When the surrounding airflow twists, the stability of the liquid jet is enhanced with the increase of the airflow rotation strength, both in the axisymmetric mode and in the asymmetry mode. According to the calculation conditions of this paper, the twist of the surrounding airflow is not conducive to the breakup of the liquid jet.

Author Contributions: Conceptualization, M.L.; methodology, M.L.; validation, L.-M.G., M.L. and Z.N.; formal analysis, M.L.; data curation, L.-M.G.; writing—original draft preparation, L.-M.G.; writing—review and editing, L.-M.G.; supervision, M.L.; project administration, M.L. and Z.N.; funding acquisition, M.L. All authors have read and agreed to the published version of the manuscript.

Funding: This project was supported by the Fundamental Research Funds for the Central Universities (grant no. 2020JBM053) and the National Natural Science Foundation of China (grant no. 51776016).

Institutional Review Board Statement: Not applicable.

Informed Consent Statement: Not applicable.

Data Availability Statement: The data that support the finding of this study are available within the article.

Conflicts of Interest: The authors declare that they have no conflict of interest.

Nomenclature

a	cylindrical nozzle radius, m
A	the gas-liquid axial velocity ratio
E	the non-dimensional rotational strength
k_i	a the disturbance spatial growth rate
$\mathrm{k_r}$	the wave number in the z direction
k_v	the bubble adiabatic index
m	the wave number in the θ direction
Ma_2	the surrounding gas Mach number
N	the number of cavitation per unit volume

$\overline{p}_1$	the liquid jet pressure, N/m^2
$\overline{p}_2$	the surrounding gas pressure, N/m^2
p_{in}	the gas phase inertia, N/m^2
p_σ	the surface tension, N/m^2
Q	the gas-liquid density ratio
r_v	the cavity average radius, m
R	the gas constant
Re_1	the liquid jet Reynolds number
T_1	the liquid jet temperature, K
U_1	the liquid jet velocity, m/s
U_2	the surrounding gas velocity in the z-axis direction, m/s
v_r	radial velocities, m/s
$v_{i\theta}$	circumferential velocities, m/s
v_{iz}	axial velocities, m/s
W_0	the surrounding gas rotation strength, m^2/s
We	the ratio of the surface tension to the inertial force
z_v	the bubble compressibility factor
α	bubble volume fraction
η	the perturbation on the liquid jet at the interface, m
μ_l	liquid viscosity, Pa·s
μ_v	cavitation bubble viscosity, Pa·s
$\overline{\rho}_1$	the liquid jet density, kg/m^3
$\overline{\rho}_2$	the gas density, kg/m^3
ρ_l	the liquid density, kg/m^3
ρ_v	the cavitation bubble density, kg/m^3
σ	the surface tension factor at the interface, N/m
ω_r	the disturbance temporal growth rate
ω_i	the wave frequency
Subscripts	
1	liquid jet parameters
2	airflow parameters

Appendix A

The expression of each element in matrix A.

$A_{11} = \mathrm{I'}_m(n_1);$
$A_{12} = \frac{m}{n_1}\mathrm{I}_m(n_1);$
$A_{13} = \frac{k}{\mathrm{i}k-\omega}\mathrm{I'}_m(k);$
$A_{14} = 0;$
$A_{15} = \mathrm{i}k - \omega;$
$A_{21} = A_{22} = A_{23} = 0;$
$A_{24} = n_2\mathrm{I'}_m(n_2);$
$A_{25} = \omega + \mathrm{i}mE - \mathrm{i}kA;$
$A_{31} = \frac{2n_1}{Re_1}\mathrm{I''}_m(n_1);$
$A_{32} = \frac{2m}{Re_1}\left[\mathrm{I'}_m(n_1) - \frac{\mathrm{I}_m(n_1)}{n_1}\right];$
$A_{33} = -\left[\mathrm{I}_m(k) + \frac{2k^2}{Re_1(\omega-\mathrm{i}k)}\mathrm{I''}_m(k)\right];$
$A_{34} = Q(\omega + \mathrm{i}mE - \mathrm{i}kA)\mathrm{K}_m(n_2);$
$A_{35} = We(k^2 + m^2 - 1) + QE^2;$
$A_{41} = 2\mathrm{i}m\left[\mathrm{I'}_m(n_1) - \frac{\mathrm{I}_m(n_1)}{n_1}\right];$
$A_{42} = \mathrm{i}\left[\frac{m^2}{n_1}\mathrm{I}_m(n_1) - \mathrm{I'}_m(n_1) + n_1\mathrm{I''}_m(n_1)\right];$
$A_{43} = \frac{2\mathrm{i}m}{\omega-\mathrm{i}k}\left[\mathrm{I}_m(k) - k\mathrm{I'}_m(k)\right];$
$A_{44} = A_{45} = 0;$
$A_{51} = \mathrm{i}\left(k + \frac{n_1^2}{k}\right)\mathrm{I'}_m(n_1);$
$A_{52} = \frac{\mathrm{i}km}{n_1}\mathrm{I}_m(n_1);$
$A_{53} = \frac{2\mathrm{i}k^2}{\mathrm{i}k-\omega}\mathrm{I'}_m(k);$
$A_{54} = A_{55} = 0$

References

1. Li, H.S.; Kelly, R.E. The instability of a liquid jet in a compressible air stream. *Phys. Fluids* **1992**, *4*, 2162–2168. [CrossRef]
2. Zhou, Z.W.; Lin, S.P. Effects of compressibility on the atomization of liquid jets. *J. Propuls. Power* **1992**, *8*, 736–740. [CrossRef]
3. Lasheras, J.C.; Hopfinger, E.J. Liquid jet instability and atomization in a coaxial gas stream. *Ann. Rev. Fluid Mech.* **2000**, *32*, 275–308. [CrossRef]
4. Gordillo, J.M.; Perez-Saborid, M.; Ganan-Calvo, A.M. Linear stability of co-flowing liquid-gas jets. *J. Fluid Mech.* **2001**, *448*, 23–51. [CrossRef]
5. Solórzano-López, J.; Zenit, R.; Ramírez-Argáez, M.A. Mathematical and physical simulation of the interaction between a gas jet and a liquid free surface. *Appl. Math. Model.* **2011**, *35*, 4991–5005. [CrossRef]
6. Zarei, M.; Davarpanah, A.; Mokhtarian, N.; Farahbod, F. Integrated feasibility experimental investigation of hydrodynamic, geometrical and, operational characterization of methanol conversion to formaldehyde. *Energy Sources Part A-Recover. Util. Environ. Eff.* **2020**, *42*, 89–103. [CrossRef]
7. Davarpanah, A.; Zarei, M.; Valizadeh, K.; Mirshekari, B. CFD design and simulation of ethylene dichloride (EDC) thermal cracking reactor. *Energy Sources Part A: Recover. Util. Environ. Eff.* **2019**, *41*, 1573–1587. [CrossRef]
8. Daryayehsalameh, B.; Nabavi, M.; Vaferi, B. Modeling of CO_2 capture ability of [Bmim][BF4] ionic liquid using connectionist smart paradigms. *Environ. Technol. Innov.* **2021**, *22*, 101484. [CrossRef]
9. Liao, Y.; Jeng, S.M.; Jog, M.A. Instability of an annular liquid sheet surrounded by swirling airstreams. *AIAA J.* **2000**, *38*, 453–460. [CrossRef]
10. Ibrahim, A.A.; Jog, M.A. Effect of liquid and air swirl strength and relative rotational direction on the instability of an annular liquid sheet. *Acta Mech.* **2006**, *186*, 113–133. [CrossRef]
11. Jog, M.A.; Ibrahim, A.A. Nonlinear breakup of a coaxial liquid jet in a swirling gas stream. *Phys. Fluids* **2006**, *18*, 14101.
12. Lin, Y.J.; Yang, Y.X.; Xi, D.G.; Qing, D. A study on the instability of an annular swirling liquid jet. *J. Eng. Thermophys.* **2001**, *22*, 519–522.
13. Kumar, A.; Sahu, S. Large scale instabilities in coaxial air-water jets with annular air swirl. *Phys. Fluids* **2019**, *31*, 124103. [CrossRef]
14. Du, Q.; Li, X.G.; Liu, N. Effect of gas swirl on breakup scale of annular liquid jet for Para-Sinuous disturbances. *J. Tianjin Univ.* **2008**, *41*, 569–575.
15. Guo, J.P.; Bai, F.Q.; Chang, Q.; Du, Q. Investigation on Asymmetric Instability of Cylindrical Power-Law Liquid Jets. *Energies* **2019**, *12*, 2785. [CrossRef]
16. Strasser, W.; Battaglia, F. Identification of Pulsation Mechanism in a Transonic Three-Stream Airblast Injector. *J. Fluids Eng.* **2016**, *138*, 111303. [CrossRef]
17. Lü, M.; Ning, Z.; Lu, M.; Sun, C. On the spatial stability of a liquid jet in the presence of vapor cavities. *Phys. Fluids* **2013**, *25*, 114107. [CrossRef]
18. Lü, M.; Ning, Z.; Yan, K.; Sun, C. Temporal and spatial stability of liquid jet containing cavitation bubbles in coaxial swirling compressible flow. *Meccanica* **2016**, *51*, 2121–2133. [CrossRef]
19. Wang, X.T.; Ning, Z.; Lü, M.; Sun, C.H. Temporal analysis of breakup for a power law liquid jet in a swirling gas. *Meccanica* **2018**, *53*, 2067–2078. [CrossRef]
20. Wang, X.T.; Ning, Z.; Lü, M. Temporal instability for a charged power-law liquid jet in a coaxial swirling air. *AIAA J.* **2018**, *56*, 3515–3523. [CrossRef]
21. Liang, X.; Deng, D.S.; Nave, J.C.; Johnson, S.G. Linear stability analysis of capillary instabilities for concentric cylindrical shells. *J. Fluid Mech.* **2011**, *683*, 235–262. [CrossRef]
22. Valizadeh, K.; Farahbakhsh, S.; Bateni, A.; Zargarian, A.; Davarpanah, A.; Alizadeh, A.; Zarei, M. A parametric study to simulate the non-Newtonian turbulent flow in spiral tubes. *Energy Sci. Eng.* **2020**, *8*, 134–149. [CrossRef]
23. Chandrasekhar, S. *Hydrodynamic and Hydromagnetic Stability*; Oxford University Press: Oxford, UK, 1961.
24. Decent, S.P.; King, A.C.; Simmons, M.J.H.; Părău, E.I.; Wallwork, I.M.; Gurney, C.J.; Uddin, J. The trajectory and stability of a spiralling liquid jet: Viscous theory. *Appl. Math. Model.* **2009**, *33*, 4283–4302. [CrossRef]
25. Lin, S.P.; Lian, Z.W. Mechanisms of the breakup of liquid jets. *AIAA J.* **1990**, *28*, 120–126. [CrossRef]
26. Yan, C.J. Atomization Mechanisms of 3-D viscous Liquid jets in a compressible gas. *Trans. CSICE* **2007**, *25*, 346–351.
27. Yan, C.J.; Xie, M.Z. Boundary conditions and Instability analysis for disturbance governing equation of liquid jets. *Chin. J. Eng. Math.* **2008**, *25*, 563–566.
28. Potter, M.C.; Wiggert, D.C. *Mechanics of Fluids*, 3rd ed.; Cengage Learning: Stamford, CT, USA, 2009.
29. Lin, S.P. *Breakup of Liquid Sheets and Jets*; Cambridge University Press: Cambridge, UK, 2003.
30. Lü, M.; Ning, Z.; Sun, C. Study on the stability of liquid jet in coaxial swirling compressible gas and cavitation bubble within a single droplet. *CSTAM J.* **2016**, *48*, 857–866.

processes

Communication

Upgrading of Wash Oil through Reduction of Nitrogen-Containing Compounds

Su Jin Kim

Department of Chemical Engineering, Chungwoon University, Incheon 22100, Korea; sujkim@chungwoon.ac.kr

Abstract: As part of improving the quality of wash oil, the reduction of three kinds of nitrogen-containing compounds (NCs), including quinoline (QU), iso-quinoline (IQU), and indole (IN), found in wash oil was examined by liquid–liquid equilibrium extraction. The wash oil and an aqueous solution of formamide were used as the raw material and the solvent, respectively. Increasing the volume fraction of water in the solvent in the initial state ($y_{w,0}$) resulted in a decrease in the distribution coefficients of each NC, while inversely, the selectivity of each NC in reference to 2-methylnaphthalene increased. The selectivity ranges of QU, IQU and IN at $y_{w,0}$ = 0.05~0.3 were 19~57, 19~56 and 50~79, respectively. Through five stages of equilibrium extraction performed under the condition of $y_{w,0}$ = 0.05, the concentrations of QU and IQU contained in the raffinate oil recovered at the fifth stage were reduced by about 69% and 65% compared to those contained in the wash oil. The concentration of IN in particular, a useful compound in the chemical industry, which is a raw material for pharmaceuticals, dyes, and fragrances, was reduced by 93.4% through a five-stage extraction operation. The formamide extraction method of this study was highly efficient in reducing the NC present in the wash oil, showing the feasibility of this method.

Keywords: coal tar; wash oil; nitrogen-containing compounds; indole; formamide equilibrium extraction

Citation: Kim, S.J. Upgrading of Wash Oil through Reduction of Nitrogen-Containing Compounds. *Processes* **2021**, *9*, 1869. https://doi.org/10.3390/pr9111869

Academic Editor: Jean Claude Assaf

Received: 22 September 2021
Accepted: 18 October 2021
Published: 20 October 2021

1. Introduction

Wash oil produced from a distillation fraction (distilled temperature 500–550 K) of coal tar contains a number of valuable nitrogen-containing compounds (NCs), such as a quinoline (QU), iso-quinoline (IQU), quinaldine and indole (IN), which are used in the production of pharmaceuticals, perfumes, pigments, dyes and pesticides [1–3]. Conversely, the NCs mentioned above are undesirable impurities in wash oil due to their contribution to air pollution and unpleasant odors. At present, QU and mixtures containing more than 0.1% QU are classified as hazardous chemicals (carcinogens, skin irritants, etc.). Therefore, wash oil containing about 1.4% QU is a hazardous chemical, so there are many restrictions on its transportation, handling, etc. For the above reasons, the improvement of wash oil quality according to the reduction of NCs and the review of the purification of useful components, such as IN, using a crude NC mixture recovered through the quality improvement process are considered to be very meaningful from the viewpoint of recycling resources.

To date, since the crude separation of NC from the coal tar fraction is a reaction extraction using an acid and a base, the operation is complicated, and has the disadvantage that it is impossible to recover and reuse the solvent [4]; thus, the development of a simpler and more efficient separation method is required. Extraction using ionic liquids [5–7], supercritical extraction [8,9], azeotropic distillation and traditional extraction methods [1–4,10–13] using various organic solvents, such as methanol, ethanol and formamide, have been studied for the reduction of NCs contained in the coal tar fraction. Among them [4,11,13], it has been reported that NCs present in the coal tar fraction can be reduced in a convenient and efficient manner by using an aqueous solution of methanol as a conventional extraction method. However, this methanol extraction method is not satisfactory from a separation point of view because of the low selectivity of NCs.

In a previous study [3], we examined the effects of experimental factors on the reduction of NCs using eight kinds of traditional extraction solvents and a model coal tar fraction consisting of nine kinds of components. In this review, it was reported that the aqueous solution of formamide is the optimal solvent in terms of the balance between the distribution coefficient and the selectivity of NCs.

This study was conducted to improve the quality of wash oil by reducing the NCs contained within it. In this review, we investigated the effect of the volume fraction of water in the solvent in the initial state ($y_{w,0}$) on the reduction of NCs present in wash oil using the aqueous solution formamide. In addition, to confirm the efficacy of reducing NCs through the batch multistage extraction under fixed extraction conditions, we performed a five-stage equilibration extraction.

2. Materials and Experimental Methods

2.1. Materials

As a raw material of this study, wash oil was supplied by OCI Company Ltd. (Seoul, Korea). Naphthalene (NA, 99% purity), QU (98% purity), IQU (97% purity), IN (≥99% purity), quinaldine (≥95% purity), 2-methylnaphthalene (2MNA, 97% purity), 1-methylnaphthalene (≥95% purity), biphenyl (BP, ≥99.5% purity), dibenzofuran (DBF, 98% purity) and fluorene (98% purity), which were used for the GC identification or quantification of the constitutive components of the wash oil, were purchased from Sigma-Aldrich, Seoul, Korea. Formamide (≥99.5% purity), used as a solvent in this study, was also purchased from Sigma-Aldrich, Seoul, Korea. In this study, the above-mentioned 11 kinds of commercial reagents were used without further purification.

2.2. Experimental Method

A schematic diagram for the extraction method used in this study is presented in Figure 1 [2]. To make a solvent of a certain concentration, the formamide was mixed with tap water. Erlenmeyer flasks of 1 L containing a certain amount of the feed (R_0: the wash oil, R_1–R_4: the raffinate phase recovered from each stage) and the fresh solvent (S_1–S_5) were placed in a shaking water bath maintained at the experimental temperature and vibrated for a certain time to reach a liquid–liquid equilibrium. After reaching the equilibrium, the mixture was settled for an aliquot of time, and the raffinate phase and the extract phase were separated using a 1 L separatory funnel, and then the mass of each phase was measured. The separated raffinate phase and a fresh aqueous solution of formamide were used as a raw material and a solvent at the next stage, respectively. Irrespective of the number of stages, 400 mL of the fresh solvent was charged in a 1 L Erlenmeyer flask. The raffinate phases and the extract phases recovered from the equilibrium extraction of each stage were analyzed by adding a small amount of acetone, and their compositions were determined. The analysis of two phases was carried out by a gas chromatograph (GC, Hewlett Packard Co., Houston, TX, USA, HP 6890: capillary column, HP-1 (60 mL, 0.32 mm I.D.)) equipped with flame ionization detector (FID). The analysis conditions of the samples were as follows: carrier gas, N_2; volume flow rate, 1 mL/min; injection temperature, 523 K; sample volume, 1 μL; splitting ratio, 0.025; column temperature, maintained at 383 K for 3 min, then increased at a rate of 5 K/min to 523 K, then 14 K/min to 593 K; detector temperature, 593 K.

2.3. Material Systems and Experimental Conditions

The material systems used for this study and the experimental conditions are summarized in Table 1. The wash oil (400 mL R_0), and the raffinate phases (R_1–R_4), which were recovered from each stage after reaching liquid–liquid equilibrium, were used as the raw materials. The fresh aqueous solution of formamide (400 mL S_1–S_5) was used as a solvent. The volume fraction of the solvent to a raw material in the initial state (S_1/R_0), the operating temperature (T) and the volume of the fresh solvent added to each stage were fixed and constant. The number of equilibrium extractions (n), the volume fraction of water in the solvent in the initial state ($y_{w,0}$) and the liquid–liquid contact time (t) were changed.

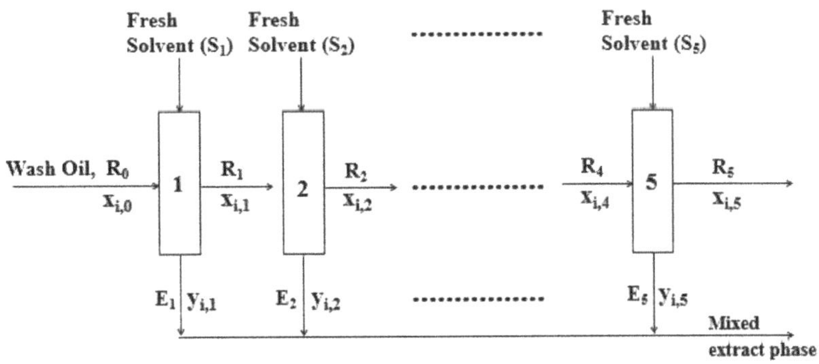

Figure 1. Schematic diagram for extraction method [2]. R: raffinate phase, E: extract phase, i: component i, 1, 2, 3, 4, 5: number of equilibrium extraction.

Table 1. Material systems and experimental conditions.

System	
Feed: wash oil and raffinate phases Solvent: aqueous solution of formamide	
Experimental Conditions	
Liquid–liquid contact time, t (h)	12~96
Number of equilibrium extractions, n (-)	1~5
Operating temperature, T (K)	303
Volume of fresh solvent (mL)	400
Volume fraction of water in solvent in initial state, $y_{w,0}$ (-)	0.05~0.3
Volume ratio of solvent to feed in initial state, S_1/R_0 (-)	1

3. Results and Discussion

3.1. Gas Chromatogram of Wash Oil

Figure 2a presents the gas chromatogram of the wash oil and the component names of the identified compounds. They were identified through analysis by adding a small amount of 10 standard reagents purchased from Sigma-Aldrich, Korea, as mentioned above. From the GC identification, we can see that the wash oil contains a relatively small amount of NC, but a large amount of bicyclic aromatic compounds (2MNA, 1-methylnaphthalene) and BP. The compositions of the six compounds quantified in this study are shown in Table 2. The compositions of QU, IQU and IN, which are the components of which this study aims to reduce the concentrations, were about 1.4%, 1.7% and 4.4%, respectively.

Table 2. Composition of feed (wash oil).

Component	Mass Fraction (-)
Naphthalene (NA, $C_{10}H_8$)	0.033
Quinoline (QU, C_9H_7N)	0.014
Iso-quinoline (IQU, C_9H_7N)	0.017
Indole (IN, C_8H_7N)	0.044
2-Methylnaphthalene (2MNA, $C_{11}H_{10}$)	0.335
Biphenyl (BP, $C_{12}H_{10}$)	0.087
Others	0.470

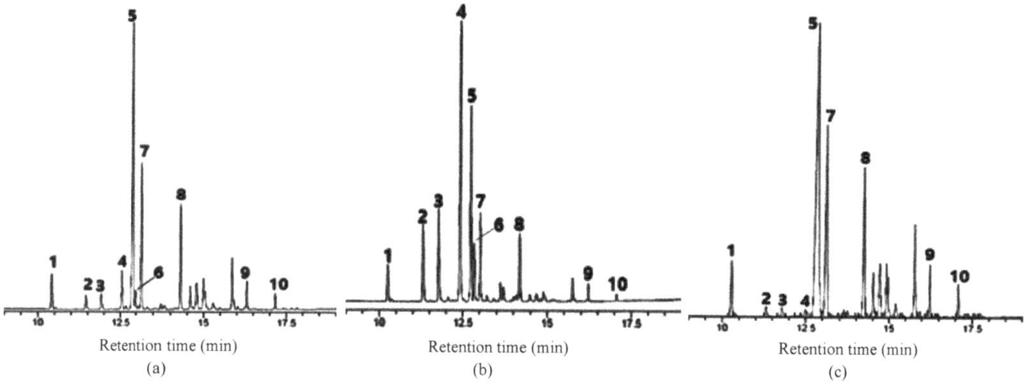

Figure 2. Gas chromatogram of (**a**) feed (wash oil), (**b**) mixed extract phase recovered through by n = 1~5 ($\sum_{n=1}^{5} E_n$) and (c) raffinate phase (R_5) recovered by n = 5. Experimental conditions: $y_{w,0}$ = 0.05, S_1/R_0 = 1, T = 303 K, t = 72 h. Peak number 1: naphthalene, 2: quinoline, 3: iso-quinoline, 4: indole, 5: 2-methylnaphthalene, 6: quinaldine, 7: 1-methylnaphthalene, 8: biphenyl, 9: dibenzofuran, 10: fluorene.

3.2. *Batch Equilibrium Extraction*

3.2.1. Definition Equation

The distribution coefficient of component i obtained at the nth stage, $(m_i)_n$, is defined as:

$$(m_i)_n = (y_i)_n/(x_i)_n \tag{1}$$

where $(y_i)_n$ and $(x_i)_n$, respectively, denote the mass fraction of component i in the extract phase and that in the raffinate phase recovered after the nth stage contact run.

The selectivity of component i in reference to 2MNA at the nth stage, $(\beta_{i,2MNA})_n$, which is defined as the ratio of the distribution coefficient for component i to that for 2MNA, was calculated by Equation (2).

$$(\beta_{i,2MNA})_n = (m_i)_n/(m_{2MNA})_n \tag{2}$$

The residual rate of component i in the raffinate phase recovered from the nth stage $(R_i)_n$ and residual rate of the raffinate phase $(R_r)_n$, respectively, are defined as:

$$(R_i)_n = R_n \times (x_i)_n/\{R_0 \times (x_i)_0\} \times 100\% \tag{3}$$

$$(R_r)_n = (R_n/R_0) \times 100\% \tag{4}$$

where R_0, R_n and $(x_i)_0$ refer to the mass of a raw material (wash oil), that of the raffinate recovered after the nth stage contact run, and the mass fraction of component i in a raw material (wash oil), respectively.

3.2.2. Review of Mass Balance and Reproducibility

As a result of calculating Equation (5) by substituting the measured values from this study to both sides of the mass balance equation, expressed as follows, the values of both sides were almost the same.

$$R_{n-1} \times (x_i)_{n-1} - R_n \times (x_i)_n = E_n \times (y_i)_n - S_n \times (y_i)_0 \tag{5}$$

where S_n and $(y_i)_0$ represent the mass of the fresh solvent put into the nth stage and the mass fraction of component i, respectively. The values of $(y_i)_0$ for all components in this study are zero, regardless of the number of the equilibrium extraction. From this result, it was estimated that the measured values through the entire experiment in this study were very satisfactory. In addition, the experiments were replicated twice or three times under the same conditions to confirm the reproducibility of the measured value. The range of mass variation of component i in the raffinate phase or the extract phase was very good, within ±5%.

3.2.3. Confirmation of Equilibrium Arrival Time

To confirm the time of reaching the equilibrium, the raffinate phases and the extract phases recovered through contact (contact time: 12 h, 24 h, 72 h, 96 h) with a raw material and a solvent were analyzed. The compositions of the raffinate phase and the extract phase recovered through contact for more than 48 h were almost the same, regardless of the contact time. Therefore, the liquid–liquid contact time of this study was maintained at 72 h.

3.2.4. Reduction Performance

Figure 3a,b, respectively, represent the effects of $y_{w,0}$ on $(m_i)_1$ and $(\beta_{i,2MNA})_1$ in the first stage (n = 1) under fixed conditions (S_1/R_0 = 1, T = 303 K, t = 72 h). Increasing $y_{w,0}$ resulted in decreasing $(m_i)_1$ because the polarity of an extract phase increases with an increase in its moisture. Therefore, $(m_i)_1$, accordingly, decreased sharply as $y_{w,0}$ increased. The $(m_i)_1$ (i = QU, IQU, IN) of three kinds of NC were much higher than those of NA, 2MNA and BP. This indicates that the polarities of three kinds of NC are much higher than those of the other components of the three kinds. As shown from the definition equation of $(\beta_{i,2MNA})_1$ and the above-mentioned results of $(m_i)_1$, the $(\beta_{i,2MNA})_1$ of three kinds of NC, inversely, increased with an increase in $y_{w,0}$. The values of $(\beta_{QU,2MNA})_1$ and $(\beta_{IQU,2MNA})_1$ were observed to be almost the same, and it was difficult to separate between those components. The sequence of $(\beta_{i,2MNA})_1$ for the entire component of this study was IN > QU = IQU > NA > BP. In $y_{w,0}$ = 0.05~0.3, the ranges of $(\beta_{i,2MNA})_1$ for QU, IQU and IN, which are the components of which this study aims to reduce the concentrations, were 19~57, 19~56, and 50~79, respectively.

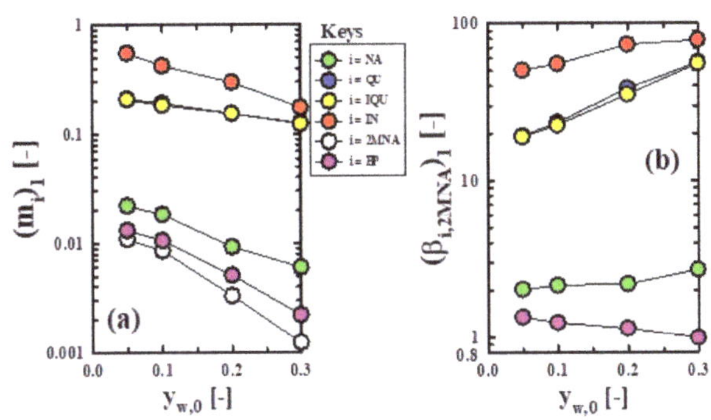

Figure 3. Changes in (**a**) $(m_i)_1$ and (**b**) $(\beta_{i,2MNA})_1$ according to $y_{w,0}$. Experimental conditions: S_1/R_0 = 1, T = 303 K, t = 72 h.

Figure 4a,b, respectively, show the changes in $(m_i)_n$ and $(\beta_{i,2MNA})_n$ according to the number of equilibrium extractions obtained through the five-stage extraction performed under $y_{w,0}$ = 0.05, selected in this study considering the balance of $(m_i)_1$ and $(\beta_{i,2MNA})_1$ for the NCs in Figure 3. Regardless of the number of equilibrium extractions, $(m_i)_n$ in all of the components in this study showed almost the same value. Therefore, we could not recognize a change in $(m_i)_n$ according to the number of equilibrium extractions. The sequence of $(m_i)_n$ was IN > QU = IQU > NA > BP > 2MNA. The same as the above-mentioned tendency of $(m_i)_n$, the $(\beta_{i,2MNA})_n$ of three kinds of NC (QU, IQU, IN) was much greater than that of other the three kinds of compounds (NA, BP, 2MNA), and $(\beta_{QU,2MNA})_n$ and $(\beta_{IQU,2MNA})_n$ were almost the same. Through this, we could see that formamide extraction is effective in reducing NCs from the wash oil, but it is difficult to separate QU and IQU. The selectivity of QU, IQU and IN were about 21, 20 and 57, respectively, regardless of the number of equilibrium extractions. It was found that the formamide extraction method used in this study showed particularly high selectivity to IN among the NC components. The sequence of selectivity for all of the components of this study was the same as that of the above-mentioned $(m_i)_n$.

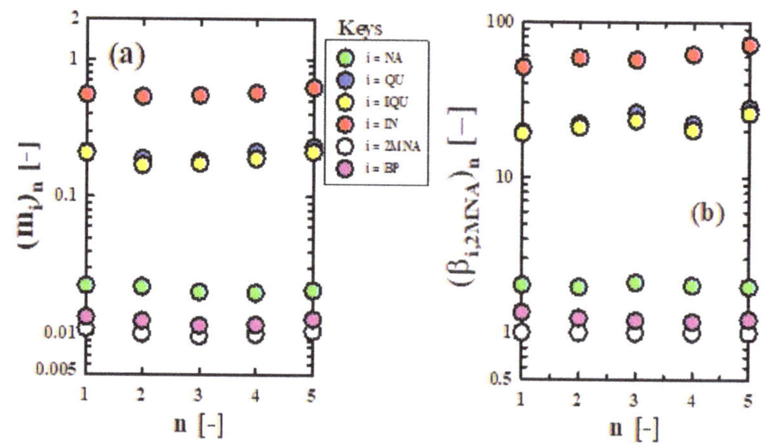

Figure 4. Changes in (**a**) $(m_i)_n$ and (**b**) $(\beta_{i,2MNA})_n$ according to number of equilibrium extractions (n). Experimental conditions: $y_{w,0} = 0.05$, $S_1/R_0 = 1$, T = 30 °C, t = 72 h.

Figure 5a–c presents the changes in $(x_i)_n$, $(R_i)_n$ and $(R_r)_n$ according to the number of equilibrium extractions, respectively. The $n = 0$ shown in these figures refers to the wash oil used as a raw material in this study. As predicted by Figure 3a, three kinds of NC with large polarity are extracted into the extraction phase, and it can be seen that the $(x_i)_n$ of three kinds of NC in raffinate phase decrease rapidly as the number of equilibrium extractions increases. Conversely, the other three components, aside from the three kinds of NC, showed a tendency to increase slightly as the number of equilibrium extractions increases. This is because the compositions of the three kinds of NC contained in the wash oil are very low, so even if a large amount of the NC with large polarity is extracted as the equilibrium extractions proceed, the composition change in components such as 2MNA in the raffinate phase is not significantly affected. At $n = 5$, the $(x_i)_n$ of each NC, which are the target components for concentration reduction in this study, was significantly lower than that of each NC contained in the wash oil used as a raw material in this study. Through this, the effect of reducing NC in the wash oil by formamide extraction could be reconfirmed. Conversely, it can be seen from Figure 5b that the $(R_i)_n$ of the three kinds of NC present in the raffinate oil decreases rapidly as the equilibrium extraction progresses, but the $(R_i)_n$ of the other three kinds of components decreases very slowly. The $(R_i)_n$ of QU, IQU and IN at $n = 5$ were about 31%, 35% and 6.6%, respectively. Considering the $(R_i)_n$ of QU, IQU and IN obtained from a five-stage equilibrium, it is expected that the raffinate oil containing little amounts of NC will be recovered when the extraction operation is performed using a tower-type multistage apparatus under optimal extraction conditions. As the equilibrium extraction proceeds from Figure 5c, it can be seen that $(R_r)_n$ decreases, indicating about 86% in $n = 5$. It was found in particular that the formamide extraction method, which has a very high selectivity for NC compared to other components, presented a very high residual rate of the product (raffinate oil) with improved quality.

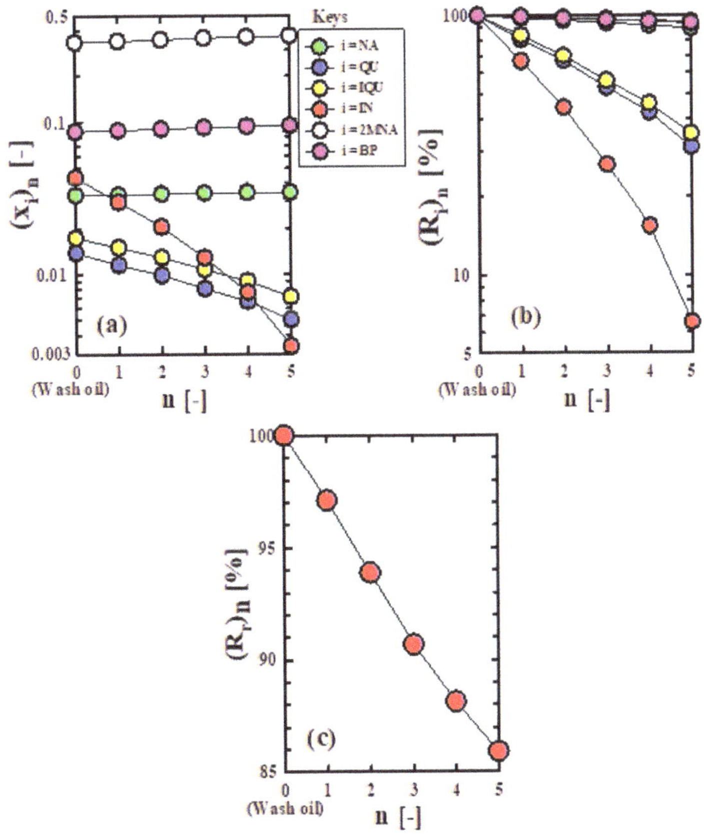

Figure 5. Changes in (**a**) $(x_i)_n$, (**b**) $(R_i)_n$, and (**c**) $(R)_n$ according to number of equilibrium extractions (n). Experimental conditions: $y_{w,0} = 0.05$, $S_1/R_0 = 1$, T = 303 K, t = 72 h.

3.2.5. Comparison of Gas Chromatogram

Figure 2b,c, respectively, show the gas chromatograms of a mixed extract phase ($\sum_{n=1}^{5} E_n$) and the raffinate phase (R_5) recovered from $n = 5$. When the gas chromatogram of the mixed extract phase in Figure 2b was compared with that of the wash oil in Figure 2a, the peak heights of four kinds of NC (peak numbers 2, 3, 4, 6) were considerably increased, but those of the other compounds were considerably decreased. The gas chromatogram of the raffinate phase (R_5) of Figure 2c was also very different from that of the wash oil. Four kinds of NC were extracted through a five-stage extraction, so that the peak height of each NC decreased, but the peak heights of other compounds were much higher. Through these gas chromatograms, it was possible to reconfirm the effect of formamide extraction on the reduction of NCs present in the wash oil.

When comprehensively considering the results of this study, as described above, the formamide extraction method is expected to be one of the most useful methods in reducing NCs in the wash oil.

3.2.6. Reduction Process of NC from Wash Oil

The process for the reduction of NCs from wash oil was investigated using the experimental results obtained from the formamide extraction [2]. The proposed process is composed of one extraction tower, one washing tower and two distillation towers, as shown in Figure 6. Tower 1 is an extraction column to separate NCs from all other compounds in the wash oil. Tower 2 is an atmospheric distillation column to remove water contained in the extract phase and tower 3 is a vacuum distillation column to separate formamide and the NC. Additionally, tower 4 is a washing column to remove the formamide present in the raffinate phase.

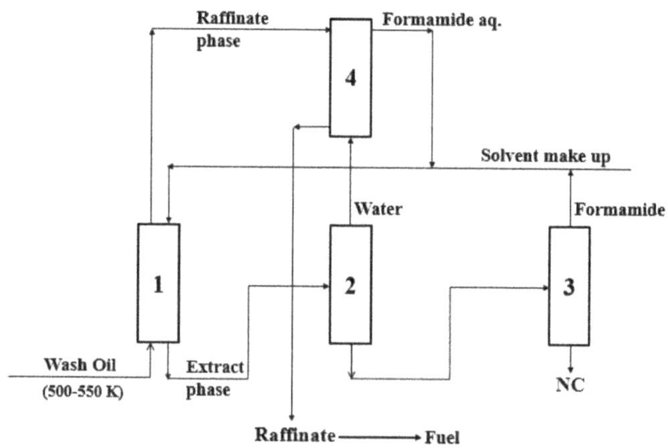

Figure 6. Reduction process of NCs contained in wash oil [2]. Tower no. 1: extraction tower, tower nos. 2, 3: distillation towers, tower no. 4: washing tower.

4. Conclusions

Until now, complex reaction extraction methods have been used to reduce NCs from a coal tar fraction, but today, the development of a simpler and more efficient reduction method is needed. From this perspective, this study experimentally performed a reduction of NCs contained in wash oil by applying a batch equilibrium extraction. The formamide extraction method, a quality improvement method of wash oil applied in this study, produced very effective results in reducing the NCs of wash oil, and it is expected to be an alternative to the complex reaction extraction methods that have been applied thus far.

Funding: The research received no external funding.

Institutional Review Board Statement: Not applicable.

Informed Consent Statement: Not applicable.

Data Availability Statement: The data presented in this study are available within the article (tables and figures).

Conflicts of Interest: The author declares no conflict of interest.

References

1. Kim, S.J. Separation and Purification of Indole in Model Coal Tar Fraction of 9 Compounds System. *Polycycl. Aromat. Compd.* **2019**, *39*, 60–72. [CrossRef]
2. Kang, H.-C.; Kim, S.J. Comparison of Methanol with Formamide on Separation of Nitrogen Heterocyclic Compounds from Model Coal Tar Fraction by Batch Cocurrent Multistage Equilibrium Extraction. *Polycycl. Aromat. Compd.* **2016**, *36*, 745–757. [CrossRef]
3. Kim, S.J.; Chun, Y.J. Separation of Nitrogen Heterocyclic Compounds from Model Coal Tar Fraction by Solvent Extraction. *Sep. Sci. Technol.* **2005**, *40*, 2095–2109. [CrossRef]
4. Egashira, R.; Nagai, M. Separation of Nitrogen Heterocyclic Compounds Contained in Coal Tar Absorption Oil Fraction by Solvent Extraction. *J. Jpn. Pet. Inst.* **2000**, *43*, 339–345. [CrossRef]
5. Xu, D.; Zhang, M.; Gao, J.; Zhang, L.; Zhou, S.; Wang, Y. Separation of heterocyclic nitrogen compounds from coal tar fractions via ionic liquids: COSMO-SAC screening and experimental study. *Chem. Eng. Commun.* **2019**, *206*, 1199–1217. [CrossRef]
6. Jiao, T.; Zhuang, X.; He, H.; Zhao, L.; Li, C.; Chen, H.; Zhang, S. An ionic liquid extraction process for the separation of indole from wash oil. *Green Chem.* **2015**, *17*, 3783–3790. [CrossRef]

7. Zhang, L.; Xu, D.; Gao, J.; Zhou, S.; Zhao, L.; Zhang, Z. Extraction and mechanism for the separation of neutral N-compounds from coal tar by ionic liquids. *Fuel* **2017**, *194*, 27–35. [CrossRef]
8. Sakanishi, K.; Obata, H.; Mochida, I.; Sakaki, T. Capture and Recovery of indole from methylnaphthalene oil in a continuous supercritical CO_2 extraction apparatus over a fixed bed of anion-exchange resin. *Ind. Eng. Chem. Res.* **1996**, *35*, 335–337. [CrossRef]
9. Sakanishi, K.; Obata, H.; Mochida, I.; Sakaki, T. Removal and Recovery of quinoline bases from methylnaphthalene oil in a semicontinuous supercritical CO_2 separation apparatus with a fixed bed of supported aluminum sulfate. *Ind. Eng. Chem. Res.* **1995**, *34*, 4118–4121. [CrossRef]
10. Uemasu, I. Effect of methanol-water mixture solvent on concentration of indole in coal tar using β cyclodextrin as complexing agent. *J. Jpn. Pet. Inst.* **1991**, *34*, 371–374. [CrossRef]
11. Ukegawa, K.; Matsumura, A.; Kodera, Y.; Kondo, T.; Nakayama, T.; Tanabe, H.; Yoshida, S.; Mito, Y. Solvent extraction of nitrogen compounds from a coal tar fraction. (Part 1). Effect of extraction conditions on the extraction rate and the selectivities of nitrogen compounds. *J. Jpn. Pet. Inst.* **1990**, *33*, 250–254. [CrossRef]
12. Egashira, R.; Salim, C. Solvent Extraction of nitrogen heterocyclic compounds contained in coal tar absorption oil fraction. improvement of separation performance by addition of aluminum chloride to solvent. *J. Jpn. Pet. Inst.* **2001**, *44*, 178–182. [CrossRef]
13. Kodera, Y.; Ukegawa, K.; Mito, Y.; Komoto, M.; Ishikawa, E.; Nakayama, T. Solvent extraction of nitrogen compounds from coal liquids. *Fuel* **1991**, *70*, 765–769. [CrossRef]

processes

MDPI

Article

An Intensified Green Process for the Coproduction of DMC and DMO by the Oxidative Carbonylation of Methanol

Abdulrahman A. Al-Rabiah [1,*], Abdulaziz M. Almutlaq [1], Omar S. Bashth [2], Taher M. Alyasser [1], Fayez A. Alshehri [1], Mohammed S. Alofai [1] and Abdulelah S. Alshehri [1,3]

1 Chemical Engineering Department, College of Engineering, King Saud University, Riyadh 11421, Saudi Arabia
2 School of Biomedical Engineering, University of British Columbia, Vancouver, BC V6T 1Z3, Canada
3 Robert Frederick Smith School of Chemical and Biomolecular Engineering, Cornell University, Ithaca, NY 14853, USA
* Correspondence: arabiah@ksu.edu.sa; Tel.: +966-11-4676844; Fax: +966-11-4678770

Abstract: Dimethyl carbonate (DMC) is an eco-friendly and sustainable compound with widespread industrial applications. Various extensive routes have been exploited in the chemical industry to produce DMC. However, these routes have several environmental and energy drawbacks. In this study, a promising novel industrial scheme for the synthesis of DMC via the oxidative carbonylation of vaporized methanol with dimethyl oxalate (DMO) as a byproduct is investigated. A methanol conversion of 81.86% and a DMC selectivity of 83.47% were achieved using an isothermal fixed-bed reactor at 130 °C. The DMC is withdrawn at a purity of >99 mol% via pressure-swing azeotropic distillations. Heat integration was performed to optimize energy consumption, reducing the energy requirements by 28%. An economic evaluation was performed for estimating the profitability via cash-flow diagrams, predicting a payback period of 3.7 years. The proposed green process exhibits several benefits, including high profitability and being environmentally friendly. It also eliminates the use or production of hazardous materials, and it enhances safety characteristics.

Keywords: dimethyl carbonate; dimethyl oxalate; oxidative carbonylation; azeotropic distillation; heat integration; profitability analysis

Citation: Al-Rabiah, A.A.; Almutlaq, A.M.; Bashth, O.S.; Alyasser, T.M.; Alshehri, F.A.; Alofai, M.S.; Alshehri, A.S. An Intensified Green Process for the Coproduction of DMC and DMO by the Oxidative Carbonylation of Methanol. *Processes* **2022**, *10*, 2094. https://doi.org/10.3390/pr10102094

Academic Editor: Jean-Claude Assaf

Received: 20 September 2022
Accepted: 13 October 2022
Published: 16 October 2022

1. Introduction

Increasing attention has been paid to the sustainable development of new environmentally benign chemicals for replacing widely used toxic reagents to alleviate the complications of harmful exposure and waste [1,2]. Dimethyl carbonate (DMC) is a promising eco-friendly chemical with a wide range of industrial applications [3]. With flammability as its sole hazard, the use of DMC eliminates the complications and precautions associated with the highly toxic phosgene and dimethyl sulfate and the carcinogenic methyl halides [4,5].

Given the eco-friendly properties and extensive applications of DMC, its demand has led to rapid annual growth, and the industry failed to satisfy the market needs. Much of the demand for DMC originates from the production of polycarbonate, a polymer that is mainly consumed in the medical-equipment and automotive industries [6]. Furthermore, owing to its nontoxicity, biodegradability, and physicochemical properties, DMC is viewed as a potential eco-friendly fuel additive that minimizes combustion-generated pollutants by inhibiting soot formation in engines [7]. Despite the commercial unavailability of DMC as a fuel additive, it exhibited similar effects to those of the oxygenate methyl tert-butyl ether (MTBE) for improving the octane performance while reducing the harmful emissions by >50% [6]. Additionally, the projected growth in demand extends to the use of DMC as a chemical reagent in methylation or carbonylation processes and as an ecofriendly electrolyte solvent in different energy-storage devices, such as high-power density double-layer capacitors and lithium batteries [8,9].

Various viable routes have been commercially exploited for synthesizing DMC using a wide range of technologies and raw materials. Traditionally, DMC was produced via the phosgenation of methanol (MeOH), which involves phosgene—a hypertoxic raw material. Phosgene is used industrially as a reagent and an intermediate for producing various materials, such as polyurethane12. However, its toxicological effects, mainly on the lungs and pulmonary system, led to its classification as a class (A) poison by the US Department of Transportation [10,11]. In spite of the high yield of the phosgene route, researchers have been working to develop inherently safer alternative routes, such as the methanolysis of urea, to mitigate the risks associated with the phosgenation production method [12]. The transesterification of both urea and ethylene carbonate (EC), along with the carbonylation of MeOH, has also been established as alternative pathways to produce DMC [13]. At present, the transesterification routes are the main industrial production methods for DMC. However, the high cost of feedstock for such routes limits the use of DMC as an eco-friendly fuel additive. Hence, the development of a profitable, sustainable, and safe route is paramount for unlocking the potential of the green compound for widespread applications [6].

Homogeneous catalysts such as cuprous chloride can be used to produce DMC in a slurry phase. However, such a route renders the separation of the catalyst and product difficult and energy-intensive [14]. To overcome the problems of homogeneous catalysts, a gas-phase oxidative carbonylation route was introduced by Curnutt and Mich [15]. A carbon-supported cupric chloride heterogeneous catalyst was used for this one-step gas-phase route to produce the DMC. The gas phase one-step process is economically more attractive when compared with the slurry phase process. There are many heterogeneous catalysts that have been investigated in the literature, and most of them are cu-based catalysts. For instance, Cu_2O, Cu/SAC, and Cu/Y-zeolite catalysts were tested in the lab for the DMC synthesis [16–18].

Herein, a novel process scheme for the synthesis of DMC via the oxidative carbonylation of vapor-phase MeOH in the presence of a CuCl2 catalyst is presented. Fang and Cao established the adequacy of the intrinsic double-rate kinetic scheme through experiments, variance tests, and residue analysis [14]. Although the use of Cu-based catalysts increases the complexity of the carbonylation reaction and introduces byproducts, such catalysts are favorable owing to their heterogeneity, which allows for the bypassing of the difficult separation of homogenous catalysts and the liquid-phase batch operation [14]. The byproduct formed by the foregoing catalytic route is dimethyl oxalate (DMO), which is a versatile feedstock to produce numerous chemicals [19]. DMO can be catalytically hydrogenated to produce a vast array of essential chemicals, such as ethylene glycol (EG) that is widely consumed in the manufacturing of polyester and coolant products as well as an organic solvent [20–22]. Furthermore, this novel process has the advantage of creating an industrial carbon cycle, as the feedstock can be directly derived from sustainable resources such as carbon dioxide and biomass. MeOH and carbon monoxide (CO) can be produced from carbon dioxide via the hydrogenation of formates and carbonates and the reaction with manganese, respectively [23,24].

The objective of this study is to develop a novel process for the production of DMC and DMO via the oxidative carbonylation of vaporized MeOH and CO. The technical assessment is built on experimental kinetics to prepare a conceptual design for modeling the process. The technical assessment is coupled with economic analysis for optimizing the separation sequence and energy requirements of the process. The process safety of this highly exothermic oxidative process is investigated.

2. Thermodynamics and Physical Properties

Valid thermodynamic properties are paramount in the process system. Their importance is increased by the presence of the DMC-H_2O and DMO-H_2O azeotropes in the separation section, which depends heavily on the prediction of reliable thermodynamic data [25]. The vapor–liquid equilibrium (VLE) can be accurately estimated via the group-contribution thermodynamic method (Universal Functional Activity Coefficient, UNIFAC). This method can handle the strongly non-ideal interactions between components in the system [26]. For the computation of such interactions, CHEMCAD was used to implement the UNIFAC model in this study.

The ternary diagram in Figure 1 shows the UNIFAC predictions of two binary azeotropes together with a map of the residue curves to the azeotropic nodes. The residue-curve map represents a collection of the liquid residue curves for a one-stage batch starting from different initial points. Combining the knowledge of the thermodynamic properties and the residue-curve map is essential for the flowsheet development and the synthesis of the separation train [27].

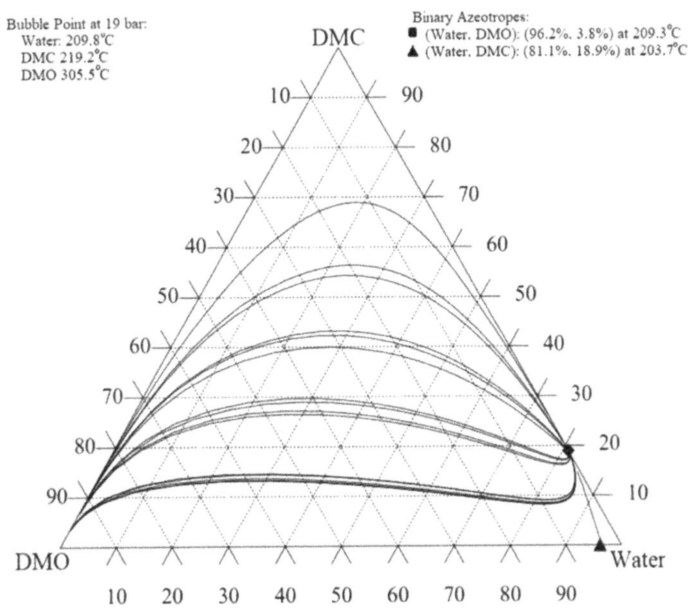

Figure 1. Ternary diagram showing the DMC-H_2O and DMO-H_2O azeotropes, along with the residue-curve map, for the distillation from the azeotropic points. The UNIFAC method was used for this calculation.

3. Reaction Kinetics and Reactor Design

3.1. Reactions and Kinetic Model

Two reactions occur on the surface of the CuCl2 catalyst: the main reaction is for the production of DMC and H_2O, and the side reaction is for the production of the byproduct, i.e., DMO, and H_2O. Both reactions consume reactants, i.e., methanol (MeOH), CO, and O_2, with different stoichiometric ratios. The two reactions are expressed as follows [14]:

$$2CH_3OH+\frac{1}{2}O_2+CO \rightarrow (CH_3O)_2CO + H_2O \quad (1)$$

$$2CH_3OH+\frac{1}{2}O_2+2CO \rightarrow (CH_3COO)_2+H_2O \quad (2)$$

The kinetic model for these two reactions was developed by Fang et al. with the help of a modified Gauss–Newton method to construct the intrinsic kinetic reaction given in Equations (3) and (4). The kinetic parameters are presented in Table 1. Variance tests and residue analysis were performed to validate the model [14].

$$r_1 = k_1.e^{\left(\frac{-E_1}{RT}\right)}.p_{1MeOH}{}^{a,1}.p_{1CO}{}^{b,1}.p_{1O_2}{}^{c,1} \tag{3}$$

$$r_2 = k_2.e^{\left(\frac{-E_2}{RT}\right)}.p_{2MeOH}{}^{a,2}.p_{2CO}{}^{b,2}.p_{2O_2}{}^{c,2} \tag{4}$$

Table 1. Parameters of reaction kinetic models [14].

Reaction No.	Constant Rate of Reaction k (mol g^{-1}h^{-1})	Activation Energy, E (J·mol^{-1})	Power Exponents a b c
1	0.3674×10^7	0.1589×10^5	1.402 0.953 0.005
2	0.1613×10^5	0.4038×10^4	0.728 1.031 0.172

An investigation of the thermodynamics of the two reactions revealed their equilibrium constants to be extremely large. Consequently, backward reactions can be safely neglected, together with any side reactions under the specified conditions. The above intrinsic reaction rates were modified in this study by introducing the effectiveness factor, which accounts for the mass transfer effect of the material inside the pores of the catalyst. Fang et al. recommended the use of a copper chloride (CuCl2)-based catalyst with activated carbon (AC1) as the first support and heteropoly acid as the second support [13]. The DMC selectivity was maximized when the reaction was conducted at a temperature of 130 °C and a pressure of 2 MPa [14].

3.2. Fixed-Bed Reactor Design

As shown in Figure 2, a fixed-bed reactor (FBR) was modeled under the assumptions of steady-state continuous, isothermal, and non-isobaric operation. The oxidative carbonylation main and side reactions are highly exothermic, with reaction enthalpies of −123.6 kJ/mol for reaction (1) and −379.2 kJ/mol for reaction (2). The reactor is operated isothermally at a temperature of 130 °C and a pressure of 2 MPa. Hence, a boiler feed water coolant inside a jacket is needed to achieve the isothermal operation at 130 °C. The actual rate of reaction (r^{act}) was modified using the effectiveness factor to consider the diffusion of reactants and products inside the pores of the catalyst, as shown in Equations (5) and (6).

$$r_1{}^{act} = r_1.\eta_1 \tag{5}$$

$$r_2{}^{act} = r_2.\eta_2 \tag{6}$$

where η is the overall effectiveness factor, and r_1 and r_2 are the intrinsic reaction rates in Equations (3) and (4), respectively. The material balance equations for each component are given as follows:

$$\frac{dF_{CO}}{dV} = (-r_1 - 2r_2)(1 - \Phi)(1 - \varepsilon)\rho_c \tag{7}$$

$$\frac{dF_{O_2}}{dV} = \left(-\frac{1}{2}r_1 - \frac{1}{2}r_2\right)(1 - \Phi)(1 - \varepsilon)\rho_c \tag{8}$$

$$\frac{dF_{MeOH}}{dV} = (-2r_1 - 2r_2)(1 - \Phi)(1 - \varepsilon)\rho_c \quad (9)$$

$$\frac{dF_{H_2O}}{dV} = (r_1 + r_2)(1 - \Phi)(1 - \varepsilon)\rho_c \quad (10)$$

$$\frac{dF_{DMC}}{dV} = (r_1)(1 - \Phi)(1 - \varepsilon)\rho_c \quad (11)$$

$$\frac{dF_{DMO}}{dV} = (r_2)(1 - \Phi)(1 - \varepsilon)\rho_c \quad (12)$$

where (Φ) is the porosity of the catalyst, which was assumed to be 0.65, and (ε) is the reactor's voidage, which is estimated to be 0.85 owing to the rapid reaction inside the packed bed. A further assumption is made that the catalyst particles have spherical shapes in order to approximate the overall effectiveness. The obtained equation is as follows [25]:

$$\eta_i = \frac{1}{\varphi_i}\left[\frac{\frac{1}{\tan 3\varphi_i} - \frac{1}{3\varphi_i}}{1 + \frac{\varphi_i}{\beta_i}\left(\frac{1}{\tan 3\varphi_i} - \frac{1}{3\varphi_i}\right)}\right] \quad (13)$$

where φ_i is the Thiele modulus, and β_i is the Biot number. These terms are calculated using the following formulas [25]:

$$\beta_i = \frac{k_{c,i}\frac{D_p}{6}}{D_{e,i}} \quad (14)$$

$$\varphi_i = \frac{D_p}{6}\sqrt{\frac{k_i}{D_{e,i}}} \quad (15)$$

where D_p is the diameter of the particles, which is taken here to be 0.0007 m [14]. For the Thiele modulus evaluation and simplification, the constant rate of the reaction (ki) in Equation (15) is assumed to be first order in CO and MeOH for both reactions. $k_{c,i}$ and $D_{e,i}$ are the mass transfer coefficient and the effective diffusivity, respectively. The mass transfer coefficient is approximated using the following Thoenes–Kramers correlation [26]:

$$k_{c,i} = D_{ab} Ree^{\frac{1}{2}} Sc^{\frac{1}{3}} \frac{\gamma\,(1 - \varepsilon)}{D_{pp}\,\varepsilon} \quad (16)$$

Here, Sc is the Schmidt number, γ is the shape factor, and Ree is given by Equation (17), where Re is the Reynolds number. The effective diffusivity, $D_{e,i}$, is estimated using the following Knudsen diffusion equation [27]:

$$Ree = \frac{Re}{\gamma\,(1 - \varepsilon)} \quad (17)$$

$$D_{e,i} = \frac{\Phi D_{ab,i}}{\tau} \quad (18)$$

The tortuosity (τ) is assumed to be 3, and the mass diffusivity (D_{ab}) is calculated as a function of the pore diameter (D_{por}), Reynolds number, temperature, and molecular weight (MW), as follows [27]:

$$D_{ab,i} = \frac{D_{por}}{3}\sqrt{\frac{8RT}{\pi M_{w,j}}} \quad (19)$$

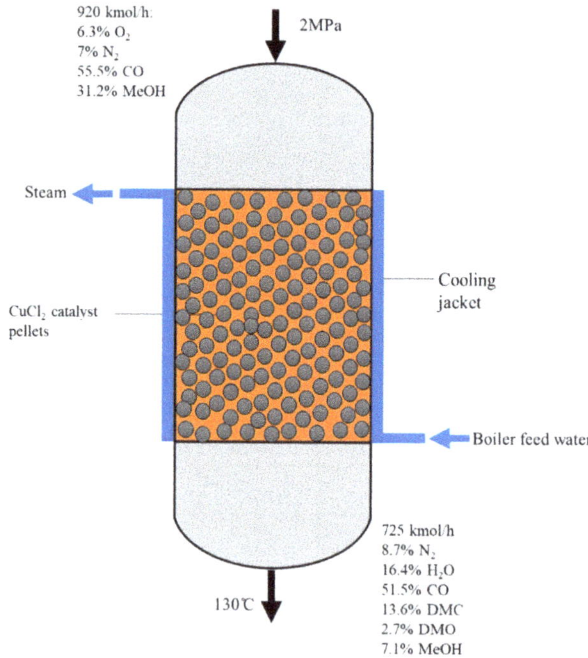

Figure 2. Schematic of the FBR and its specifications.

Since the process is non-isobaric, the momentum balance is applied for the calculation of the pressure drop inside the FBR using the following Ergun equation [28]:

$$\frac{dP}{dV} = \frac{B_o(1-\Phi)}{A_c} \cdot \frac{P_o}{P} \cdot \frac{F_T}{F_{T_O}} \tag{20}$$

where B_o is a constant that depends on the particle diameter and the reactor voidage. It is given by the following equation [28]:

$$B_o = \frac{G(1-\varepsilon)}{\rho_o g_c D_p}\left[\frac{150(1-\varepsilon)\mu}{D_p} + 1.75G\right] \tag{21}$$

Equations (5)–(21) were solved and validated using simulation software to determine the reactor volume and effluent composition. The pressure drop through the FBR was estimated to be 80 kPa. After the data points were validated using the kinetic model, the volume of the reactor needed to allow the limiting reactant, i.e., O_2, to be almost completely consumed was identified as approximately 1.3 m^3. Figure 3 displays the mole fraction of the components with respect to the reactor volume used in the process flowsheet, indicating a complete conversion of O_2. Under these conditions, the per-pass MeOH conversion and DMC selectivity were approximately 81.86% and 83.47%, respectively. The MeOH conversion and DMC selectivity were calculated as follows:

$$X_M = \frac{F_{in,\ MeOH} - F_{out,\ MeOH}}{F_{in,\ MeOH}} \tag{22}$$

$$S_{DMC} = \frac{F_{out,\ DMC} - F_{in,\ DMC}}{(F_{out,\ DMC} - F_{in,\ DMC}) + (F_{out,\ DMO} - F_{in,\ DMO})} \tag{23}$$

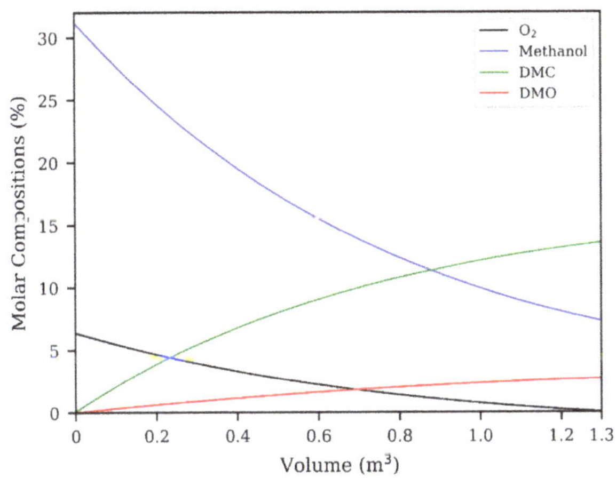

Figure 3. Molar composition of reactants and products along the reactor volume.

4. Flammability Analysis

A flammability study was performed on the reactor influent due to the reduction in flammability hazard as the reactions consume oxygen to produce DMC and DMO, leaving only traces of oxygen in the reactor outlet stream. The fuel mixture in the diagram solely consists of MeOH and CO, as shown in Table 2. The table also provides the lower and upper flammability levels at the standard temperature and pressure (25 °C and 0.101 MPa, respectively) and at 130 °C and 2 MPa for each component and the fuel mixture. The analysis was conducted using the empirical equations (Equations (24)–(26)), which provide estimates of the effects of the temperature and pressure on the flammability limits [29]:

$$\mathrm{LFL(T)} = \mathrm{LFL}(25\ ^\circ\mathrm{C}) - \frac{0.75}{\Delta \mathrm{H_c}}(\mathrm{T} - 25) \tag{24}$$

$$\mathrm{UFL(T)} = \mathrm{UFL}(25\ ^\circ\mathrm{C}) - \frac{0.75}{\Delta \mathrm{H_c}}(\mathrm{T} - 25) \tag{25}$$

$$\mathrm{UFL(P)} = \mathrm{UFL}(25\ ^\circ\mathrm{C}) + 20.5\ (\log \mathrm{P} + 1) \tag{26}$$

where LFL(T) is the lower flammability limit (vol. %), UFL(T) is the upper flammability limit (vol. %), T is the temperature (K), ΔH_c is the net heat of combustion (kcal/mol), and P is P is the pressure (MPa).

Table 2. Results of flammability analysis for reactor influent.

Component	Vol.%	Yi	LFL% at 25 °C	UFL% at 25 °C	LFL% at 130 °C	UFL% at 130 °C	LFL% mix.	UFL% mix.
MeOH	31.21	0.36	7.3	36.00	6.74	36.56	-	-
CO	55.52	0.64	12.5	74.00	11.34	75.16	-	-
O_2	6.37	-	-	-	-	-	-	-
N_2	6.90	-	-	-	-	-	-	-
Total	100	-	-	-	-	-	9.10	83.62

After the predictions of the flammability limits for the different components were calculated, a correlation for approximating the flammability levels of mixtures was applied, as follows [30]:

$$\mathrm{LFL}_{mix} = \frac{1}{\sum_{i=1}^{n}\left(\frac{Y_i}{LFL_i}\right)} \tag{27}$$

$$\text{UFL}_{mix} = \frac{1}{\sum_{i=1}^{n}\left(\frac{Y_i}{UFL_i}\right)} \tag{28}$$

where Y_i is the mole fraction of component i on a combustible basis, and n is the number of species.

Figure 4 shows the flammability diagram for the fuel/oxygen/nitrogen mixture that flows into the reactor at a temperature of 130 °C and 2 MPa, with the flammability zone indicated by red. In Figure 4, the top point on the right side, i.e., the operating point, represents the composition of the reactor feed, which is well above both the flammability region and the limiting oxygen concentration (LOC) line. Combustion is not possible for any fuel concentration above the LOC line. The dilution of oxygen with an inert gas (in this case, nitrogen) not only improves the reaction yield of the DMC but also affords a lean mixture by maintaining the reactor influent composition outside the flammability region.

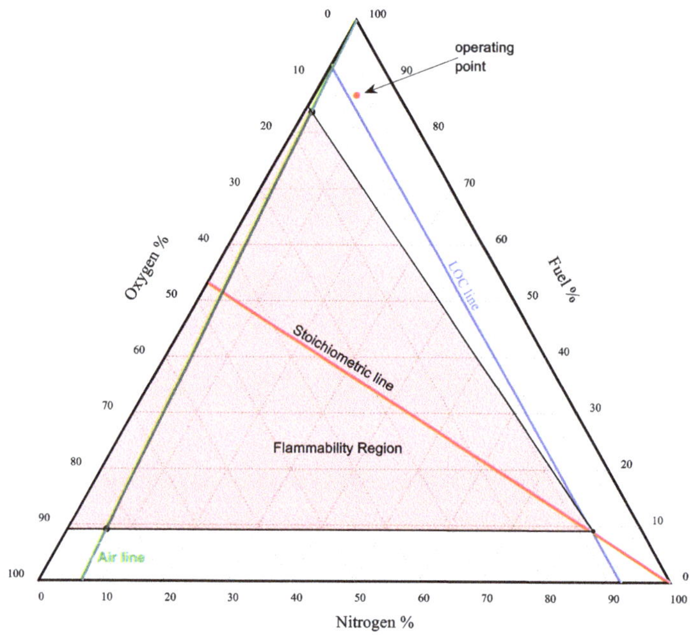

Figure 4. Fuel flammability diagram for the fuel/oxygen/nitrogen mixture at 130 °C and 2 MPa. The fuel mixture is solely composed of MeOH and CO.

5. Process Development

Under the adopted reaction route, the DMC process comprises three key sections: oxidative carbonylation, conventional distillation, and pressure-swing azeotropic distillation. The block flow diagram (BFD) represents the compilation of the three sections, as shown in Figure 5. In the first section, the reaction is conducted in the presence of a heterogeneous catalyst.

The process feed consists of vaporized MeOH, CO, O_2, and N_2, and general plant support requires power generation and cooling water. In the conventional distillation section, MeOH and DMO are separated in the two columns from the reactor effluent stream. MeOH is recycled back to the reactor, and the DMO is purified as a byproduct. The following section describes the azeotropic separation between DMC and H_2O for satisfying the target purity of DMC. The desired purities of both products—DMC and DMO—were set as >99 mol%.

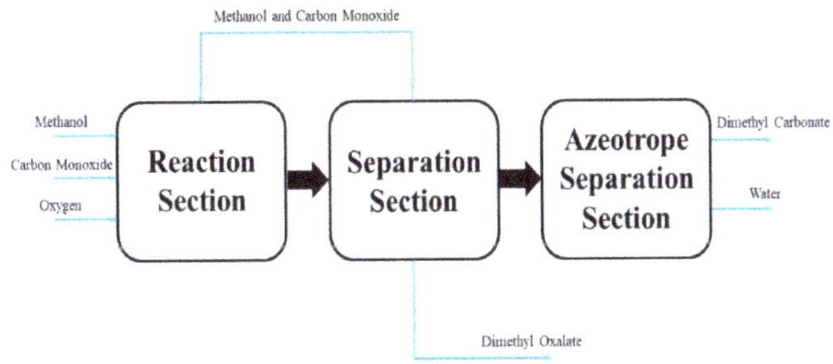

Figure 5. Block flow diagram (BFD) for the production of DMC and DMO via the oxidative carbonylation of MeOH.

5.1. Process Flowsheet Simulation

The flowsheet of the process shown in Figure 6 was developed and simulated using CHEMCAD with UNIFAC as the thermodynamic model. UNIFAC is a widely utilized thermodynamic model that exploits structural groups for estimating component interactions [31].

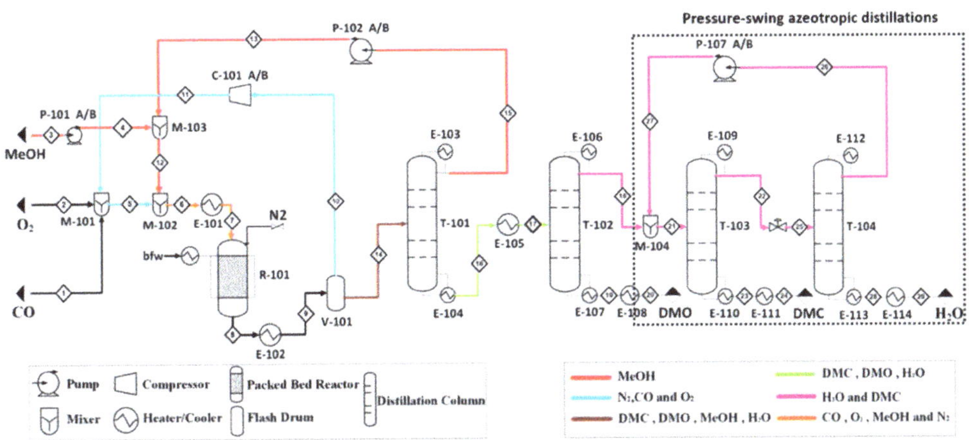

Figure 6. Process flow diagram (PFD) for the production of DMC and DMO via the oxidative carbonylation of MeOH.

The fresh feed streams are mixed with recycled MeOH and recycled gases (CO and N_2) and combined in stream 6. Prior to heating the feed, the pressure of both the liquid and gases is increased to 2 MPa using the pump (P-101) and compressor (C-101), respectively. A heat exchanger (E-101) is used to heat the feed to the desired temperature of 130 °C (403 K), in accordance with the experiment performed by Fang et al. [14].

An inert gas N_2 is essential for diluting the gaseous components, as indicated [14]. If N_2 is fed with the raw materials, a purge is needed to avoid the buildup in the vessel as the inert is not consumed in the process. However, a purge stream would consist of toxic CO. Therefore, N_2 is introduced to the reactor only once (in the beginning) and is then recycled within the process to avoid purging.

Subsequently, the reactor effluent is cooled and sent to a flash drum (V-101), where non-condensable gases are separated from products and MeOH. The non-condensable gases are recycled back and mixed with the fresh gaseous stream. The condensate stream, which contains DMC, MeOH, H_2O, and DMO, is sent to the first distillation column. MeOH is separated as an overhead product in the first distillation column (T-101) and recycled back to be mixed with the fresh feed. The bottom stream is sent to a heat exchanger (E-105) before entering the second distillation. In the second distillation column (T-102), DMO is separated in the bottom stream with a purity of 99.9 mol%. The DMO is then cooled to be sent to a storage tank.

5.2. Separation of Azeotropic Mixture

As shown in Figure 6, the pressure-swing technique is employed for the separation of the azeotropic mixture of DMC and H_2O. DMC is separated as a bottom product in the distillation column (T-103), with a purity of approximately 99.78 mol%, while H_2O is purified in the distillation column (T-104) in the bottom stream, with a purity of 99.96 mol%. The high purity of both streams eliminates the need for further purifications. The water stream is then sent to a wastewater treatment unit. The presence of azeotropes complicates the separation of mixtures by narrowing the feasible operation region of the vapor–liquid envelope.

In the first distillation column (T-101), an azeotropic mixture is formed between DMC and MeOH posing a common difficult and energy-intensive separation problem [32]. However, the availability of DMO and H_2O in the stream entering the distillation column breaks the distillation boundaries restricted by the DMC-MeOH azeotrope and allows for extractive distillation. Additionally, DMC and H_2O form an azeotropic mixture that requires the application of unconventional distillation techniques, such as pressure swing, to attain the desired separation [33].

In this process, the azeotrope between DMC and H_2O occurs at 19 bar at a DMC composition of 19 mol%, as shown in Figure 7A. To perform the separation of DMC from H_2O, the pressure must be reduced significantly (to <0.6 bar) for avoiding azeotropic separation, as illustrated in Figure 7C. However, this is impractical, as it would require vacuum distillation, which is mostly expensive to operate. The implementation of the pressure-swing technique is depicted in Figure 7. The technique allows for achieving a DMC purity of >99 mol% at the bottom of the distillation column (T-103), which is operated at 19 bar. The remaining DMC in the distillate stream (stream 22) has a mole fraction of approximately 19.3 mol%, which is slightly above the azeotropic point located around 19 mol% at 19 bar. This requires the distillation column to be followed by a valve for reducing the pressure of the mixture to a 10 bar, shifting the azeotropic point to a new composition of DMC at 20.50 mol%, as shown in Figure 7B. By shifting the azeotropic point, this pressure swing causes the DMC to act as the heavy component in column (T-103), and as the light key in the following column (T-104). The DMC- H_2O mixture is then sent to another distillation column (T-104), which separates water in the bottom stream with a high purity of approximately 99.96 mol%. The overhead of the distillation column (T-104) is recycled back to the distillation column (T-103) after its pressure is increased back to 19 bar. A full stream table of the process is provided in Table 3. The main design variables and specifications for each of the DMC plant components are presented in Table 4.

Table 3. Stream information of the DMC and DMO coproduction process.

Stream No.	1	2	3	7	9	11
Temperature (°C)	25	25	25	130	30	38.4
Pressure (bar)	1	1	1	20	19.2	20.3
Vapor mole fraction	1	1	0	1	0.6059	1
Total flow (kg/h)	3830.4	1878.8	7525.1	27,208	27,208.1	12,346.6
Total flow (kmol/h)	136.8	58.7	234.9	920.5	725	439.4

Table 3. *Cont.*

Component flow rates (kmol/h)						
O_2	0	58.7	0	58.7	0	0
N_2	0	0	0	63.6	63.6	63.6
H_2O	0	0	0	0.5	118	0.4
CO	136.8	0	0	510.2	373.4	373.4
DMC	0	0	0	0.6	98.6	0.6
DMO	0	0	0	0	19.4	0
MeOH	0	0	234.9	286.9	52	1.4
Stream No.	**12**	**13**	**14**	**17**	**18**	**20**
Temperature (°C)	55.3	164.2	30	225.2	208.3	1.5
Pressure (bar)	20.3	20.3	19.2	19.7	19.7	20
Vapor mole fraction	0	0	0	0.948	0	0
Total flow (kg/h)	9150.6	1625.5	14,861.5	13,236	10,948.4	2287.6
Total flow (kmol/h)	285.6	50.8	285.6	234.9	215.5	19.4
Component flow rates (kmol/h)						
O_2	0	0	0	0	0	0
N_2	0	0	0	0	0	0
H_2O	0.1	0.1	117.5	117.4	117.4	0
CO	0	0	0	0	0	0
DMC	0	0	98	98	98	0
DMO	0	0	19.4	19.4	0	19.4
MeOH	285.5	50.7	50.7	0	0	0
Stream No.	**21**	**22**	**24**	**27**	**29**	
Temperature (°C)	81.9	201.1	30	170.9	30	
Pressure (bar)	19	19	1.5	19	1.5	
Vapor mole fraction	0	0	0	0	0	
Total flow (kg/h)	16,895.4	8065.5	8829.8	5949.2	2116.3	
Total flow (kmol/h)	348	249.8	98.2	132.6	117.3	
Component flow rates (kmol/h)						
O_2	0	0	0	0	0	
N_2	0	0	0	0	0	
H_2O	195.6	195.4	0.2	78.2	117.3	
CO	0	0	0	0	0	
DMC	146.2	48.3	98	48.2	0	
DMO	0	0	0	0	0	
MeOH	6.2	6.2	0	6.2	0	

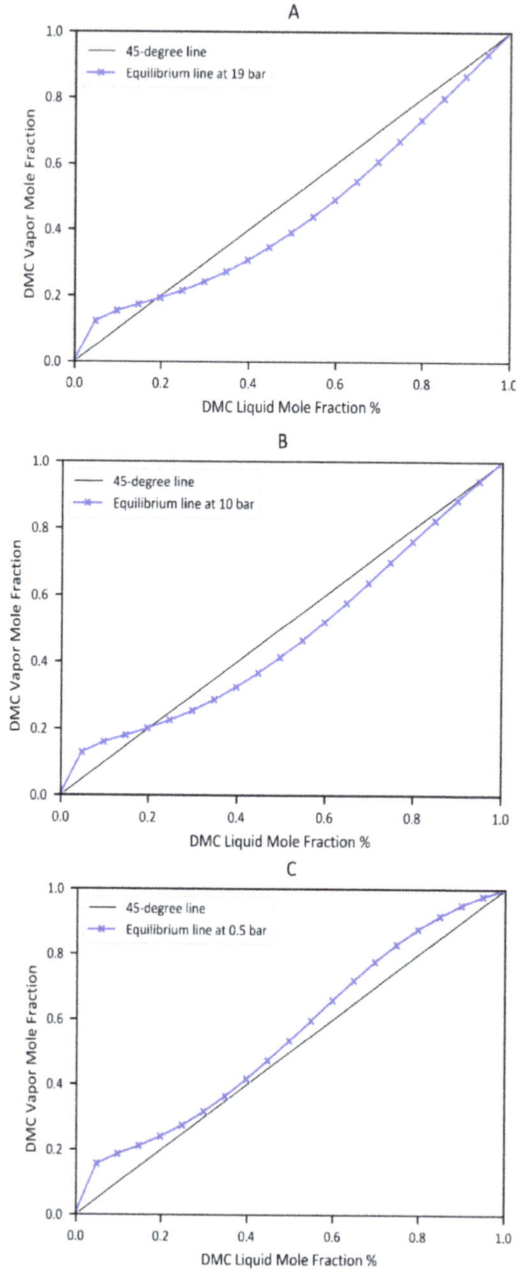

Figure 7. VLE curves of DMC with H_2O at constant pressures, obtained using the UNIFAC model. (**A–C**) presents the equilibrium at 19, 10, and 0.5 bar, respectively.

Table 4. Process main equipment design specifications.

Components	Variables	Specifications
Reactor (R-101)	Type	FBR
	Volume	1.3 m^3
	Length	1.5 m
	Diameter	0.53 m
Distillation column (T-101)	Reflux ratio	4.709
	Number of stages	44
	Condenser duty	−7508 MJ/h
	Column diameter	1.18 m
Distillation column (T-102)	Reflux ratio	0.4794
	Number of stages	28
	Condenser duty	−9326 MJ/h
	Column diameter	0.82 m
Distillation column (T-103)	Reflux ratio	6.18
	Number of stages	73
	Condenser duty	−57,863 MJ/h
	Column diameter	0.95 m
Distillation column (T-104)	Reflux ratio	0.52
	Number of stages	24
	Condenser duty	−6684 MJ/h
	Column diameter	0.6 m

6. Heat Integration

To address a new avenue for cost savings in this study, a preliminary heat integration study using the online pinch analysis tool developed by Umbach and Nitsche was investigated [34]. The pinch method involves a thermodynamic analysis of the process that determines the temperature above or below the degree of which heating and cooling utilities should be avoided in the process. The pinch temperature depends on the temperature difference between streams, as well as the flow rate of utilities and the process configuration [35].

The analysis is employed to build a network for exchanging heat between streams for minimizing the overall utility costs [36]. The underlying considerations for identifying the pinch temperature include the following: no heat passes over the pinch point, external heating input is only allowed above the pinch point, and external heating output is only permissible below the pinch point [37].

Using the online pinch analysis tool [34], the pinch temperature of the system was determined to be 213 °C (486.15 K). An allowable temperature difference (ΔT_{min}) of 10 °C. (ΔT_{min}) was measured to determine the minimum driving force allowed for the heat transfer; hence, this criterion defines the energy requirement of the process [38]. Figure 8A,B present composite curves of cold and hot streams with ΔT_{min} = 10 °C. Both curves were shifted by ±5 °C to generate the pinch point. The grand composite curve in Figure 8C indicates the minimum required heating utilities (Q_{Hmin}) and the minimum required cooling utilities (Q_{Cmin}). The implementation of heat integration delivers energy savings of up to 28%. Further optimization of ΔT_{min} can be applied to enhance the overall process integration. The utility cost can be further reduced if each stream is used more than once in stream-matching. Additionally, this preliminary heat integration does not incorporate the capital cost for the piping and heat exchanges required for the heat integration.

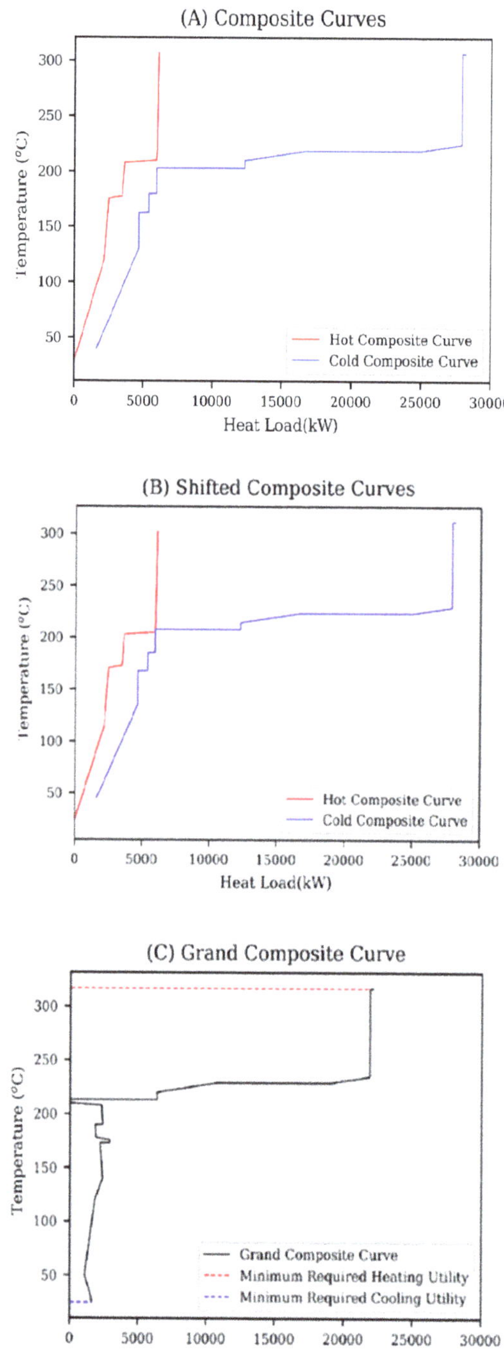

Figure 8. Heat integration curves (**A**) cold and hot composite curves showing ΔT_{min} and the process pinch temperature. (**B**) composite curves shifted by ±5 K. (**C**) grand composite curve, showing the minimum required heating utility (Q_{Hmin}) and the minimum required cooling utility (Q_{Cmin}).

7. Profitability Analysis

In examining the viability of this new DMC production process, the capital cost was combined with the operating cost to determine the overall process cost and to evaluate the financial performance over a 10-year plant lifetime. Most methods for estimating the purchase cost (C_p°) are for the ambient operating pressure, with carbon steel as the construction material. For correcting the purchase cost in this scheme, two factors (F_M and F_P) were considered for the construction materials and operating pressures. Both factors were approximated using multiple established correlations [39].

Operating-cost calculations based on 330 working days per year with 35 days of shut down for maintenance and service (yearly working hours) were performed. To include the effect of economic inflation, the Chemical Engineering Plant Cost Index (CEPCI) was applied to scale the cost with respect to time. CEPCIs of approximately 607.5 for 2019 were assumed to account for inflation [40]. The fixed capital investment of this process is about USD 10.8 million and the cost of manufacturing is about USD 51.59 million.

A 10-year profitability analysis of the proposed process was performed, with the assumption of two years of construction before the plant is operated. The fixed capital investment (FCI) was divided equally between the first and second years of construction. By the end of the construction period, the working capital cost, which was assumed to be 15% of the FCI, was added. Starting from the third year, a five-year period of depreciation of the equipment was considered using the modified accelerated cost recovery system.

For an interest rate of 7%, a discounted cumulative cash flow (DCCF) diagram with respect to time was constructed, as shown in Figure 9. The plot also shows a comparison of the profitability of the process before and after the heat integration. The discounted payback period (DPP) represents the time when the initial investment will be recovered [39]. The implementation of heat integration caused the DPP to decrease significantly from 7 years to three years and seven months, as shown in Figure 9.

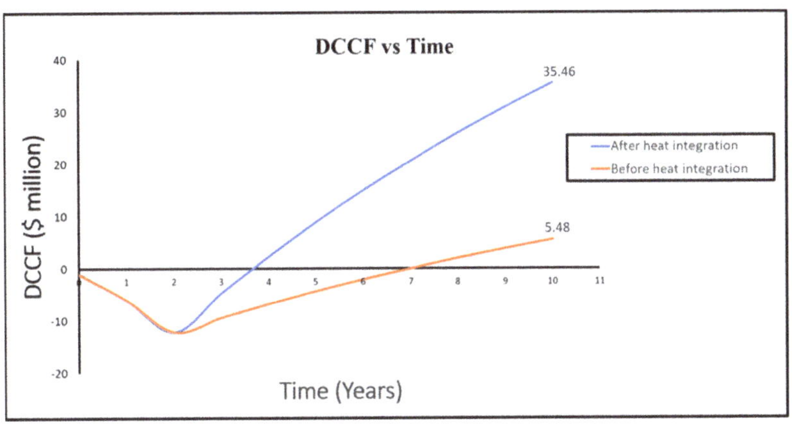

Figure 9. DCCF with respect to time in a 10-year profitability analysis.

Furthermore, the net present value (NPV) increased from USD 5.48 million to USD 35.46 million by the end of the 10th year. An analysis of the profitability before and after the heat integration is presented in Table 5.

This includes the discounted cash flow rate of return (DCFROR), which represents the interest rate at which the project would break even. Table 5 also shows the present value ratio (PVR), which is the ratio between the positive and negative discounted cash flows. The economic evaluations indicate that the DMC production process is a profitable venture, and it highlights the impact of heat integration in optimizing the process through the minimization of the operating costs.

Table 5. Process profitability analysis.

Index	Before Process Integration	After Process Integration
NPV	USD 5.48 million	USD 35.46 million
Payback period	7 years	3.7 years
DCFROR	17%	56.4%
PVR	1.4	4

8. Conclusions

A novel configuration for the production of DMC via the oxidative carbonylation of MeOH with DMO as byproducts is proposed. A techno-economic evaluation of the process was performed to assess its applicability and feasibility. The analysis results suggest that this process achieves the target purities of the final products while generating high returns on the invested capital. A process flowsheet was developed and simulated using UNIFAC as the thermodynamic model. DMC and DMO were produced on a copper chloride catalyst in an isothermal FBR, reaching a MeOH conversion rate of 81.86% and a DMC selectivity of 83.47%. DMO was purified through conventional distillation at 99.9 mol%, and a 99.78 mol% pure DMC product was obtained via the pressure-swing technique, which was employed to separate the DMC-H_2O azeotropic mixture. A profitability analysis for a 10-year plant lifetime indicated an NPV of USD 5.48 million and a payback period of seven years. To optimize the utility consumption, a preliminary heat integration was implemented, resulting in a 28% energy savings in the utilities and a reduction in the payback period to three years and seven months. The new process is considered green since it is environmentally friendly, produces a green byproduct in addition to the main product, and avoids the use of hazardous materials, as in the case of the phosgenation production method.

Author Contributions: Conceptualization, A.A.A.-R.; methodology, A.A.A.-R. and A.M.A.; software, A.A.A.-R., O.S.B., T.M.A., F.A.A. and M.S.A.; validation, A.A.A.-R., A.M.A., O.S.B., T.M.A., F.A.A. and M.S.A.; formal analysis, A.A.A.-R., A.M.A., O.S.B., T.M.A., F.A.A. and M.S.A.; investigation, A.A.A.-R., A.M.A., O.S.B., T.M.A., F.A.A. and M.S.A.; visualization, A.S.A., re-sources, A.A.A.-R.; data curation, A.A.A.-R.; writing—original draft preparation, A.A.A.-R., A.M.A., O.S.B., T.M.A., F.A.A. and M.S.A.; writing—review and editing, A.A.A.-R. and O.S.B.; visualization, A.A.A.-R. and A.S.A.; supervision, A.A.A.-R. and A.M.A.; project administration, A.A.A.-R.; funding acquisition, A.A.A.-R. All authors have read and agreed to the published version of the manuscript.

Funding: There is no external funding.

Data Availability Statement: Not applicable.

Acknowledgments: This project was supported by King Saud University, Deanship of Scientific Research, College of Engineering Research Center.

Conflicts of Interest: The authors declare no conflict of interest.

Nomenclature

A_c	Cross-sectional area of the reactor (m^2)
B_o	Constant that depends on the properties of the fixed bed (MPa/m)
C_{BM}	Bare module cost
D_p	Diameter of particles in the bed (m)
$D_{e,i}$	Effective diffusivity (m^2/s)
E	Activation energy (J/mol)
F_i	Molar flow rate for each component (kmol/h)
F_{T_o}	Total inlet flow rates (kmol/h)
F_T	Total outlet flow rate (kmol/h)

G	Superficial mass velocity $\left(kg/m^2.s\right)$
k_n	Constant rate of reaction
$k_{c,i}$	Mass transfer coefficient (m/s)
P	Final pressure (MPa)
$p_{i,n}$	Partial pressure of components (MPa)
P_o	Initial pressure (2 MPa)
r_n	Reaction rate ($mol{\cdot}g^{-1}\,h^{-1}$)
r_n^{act}	Actual rates of reaction ($mol{\cdot}g^{-1}\,h^{-1}$)
R	Universal gas constant (8.314 $J{\cdot}mol^{-1}\,K^{-1}$)
T	Reaction temperature (403.15 K)
V	Volume of the reactor (m^3)
Greek letters	
η	Overall effectiveness factor
Φ	Porosity of the catalyst
β_i	Biot number
ε	Voidage of the reactor
μ	Viscosity of the mixture (Pa·s)
ρ_c	Inlet mixture density
Indices	
i	Component
a, b, c	Power exponents of the reaction rate equations
n	Reaction number

References

1. Varma, R.S. Greener and Sustainable Chemistry. *Appl. Sci.* **2014**, *4*, 493–497. [CrossRef]
2. Crawford, S.E.; Hartung, T.; Hollert, H.; Mathes, B.; van Ravenzwaay, B.; Steger-Hartmann, T.; Studer, C.; Krug, H.F. Green Toxicology: A strategy for sustainable chemical and material development. *Environ. Sci. Eur.* **2017**, *29*, 16. [CrossRef] [PubMed]
3. Wang, F.; Pu, Y.; Yang, J.; Wang, T.; Chen, L.; Zhao, N.; Xiao, F. Process design and economic optimization for the indirect synthesis of dimethyl carbonate from urea and methanol. *Chin. J. Chem. Eng.* **2018**, *27*, 1879–1887. [CrossRef]
4. Tundo, P.; Selva, M. The Chemistry of Dimethyl Carbonate. *Accounts Chem. Res.* **2002**, *35*, 706–716. [CrossRef] [PubMed]
5. Fiorani, G.; Perosa, A.; Selva, M. Dimethyl carbonate: A versatile reagent for a sustainable valorization of renewables. *Green Chem.* **2018**, *20*, 288–322. [CrossRef]
6. Tan, H.-Z.; Wang, Z.-Q.; Xu, Z.-N.; Sun, J.; Xu, Y.-P.; Chen, Q.-S.; Chen, Y.; Guo, G.-C. Review on the synthesis of dimethyl carbonate. *Catal. Today* **2018**, *316*, 2–12. [CrossRef]
7. Saada, R.; Aboelazayem, O.; Kellici, S.; Heil, T.; Morgan, D.; Lampronti, G.I.; Saha, B. Greener synthesis of dimethyl carbonate using a novel tin-zirconia/graphene nanocomposite catalyst. *Appl. Catal. B Environ.* **2018**, *226*, 451–462. [CrossRef]
8. Ferrer, B.; Alvaro, M.; Garcia, H. *Application of Dimethyl Carbonate as Solvent and Reagent*; Springer: Cham, The Netherlands, 2012. [CrossRef]
9. Park, J.H.; Jeon, J.Y.; Lee, J.J.; Jang, Y.; Varghese, J.K.; Lee, B.Y. Preparation of High-Molecular-Weight Aliphatic Polycarbonates by Condensation Polymerization of Diols and Dimethyl Carbonate. *Macromolecules* **2013**, *46*, 3301–3308. [CrossRef]
10. Gift, J.S.; McGaughy, R.; Singh, D.V.; Sonawane, B. Health assessment of phosgene: Approaches for derivation of reference concentration. *Regul. Toxicol. Pharmacol.* **2008**, *51*, 98–107. [CrossRef] [PubMed]
11. Kongpanna, P.; Pavarajarn, V.; Gani, R.; Assabumrungrat, S. Techno-economic evaluation of different CO_2-based processes for dimethyl carbonate production. *Chem. Eng. Res. Des.* **2015**, *93*, 496–510. [CrossRef]
12. Esan, A.O.; Adeyemi, A.D.; Ganesan, S. A review on the recent application of dimethyl carbonate in sustainable biodiesel production. *J. Clean. Prod.* **2020**, *257*, 120561. [CrossRef]
13. Garcia-Herrero, I.; Cuéllar-Franca, R.M.; Enríquez-Gutiérrez, V.M.; Alvarez-Guerra, M.; Irabien, A.; Azapagic, A. Environmental Assessment of Dimethyl Carbonate Production: Comparison of a Novel Electrosynthesis Route Utilizing CO_2 with a Commercial Oxidative Carbonylation Process. *ACS Sustain. Chem. Eng.* **2016**, *4*, 2088–2097. [CrossRef]
14. Fang, D.; Cao, F. Intrinsic kinetics of direct oxidative carbonylation of vapour phase methanol to dimethyl carbonate over Cu-based catalysts. *Chem. Eng. J.* **2000**, *78*, 237–241. [CrossRef]
15. Curnutt, G.L. Catalytic vapor phase process for producing dihydrocarbyl carbonates. U.S. Patent 5004827, 2 April 1991.
16. Wang, J.; Fu, T.; Meng, F.; Zhao, D.; Chuang, S.S.; Li, Z. Highly active catalysis of methanol oxidative carbonylation over nano Cu_2O supported on micropore-rich mesoporous carbon. *Appl. Catal. B Environ.* **2021**, *303*, 120890. [CrossRef]
17. Ren, X.; Quan, Y.; Yang, W.; Zhao, J.; Shi, R.; Ren, J. Highly efficient super activated carbon supported ultra-low loading copper catalyst for the oxidative carbonylation of methanol to dimethyl carbonate. *Mol. Catal.* **2022**, *531*, 112694. [CrossRef]

18. Lee, D.-H.; You, J.; Woo, J.-M.; Seo, J.Y.; Park, Y.C.; Lee, J.-S.; Kim, H.; Moon, J.-H.; Bin Park, S. Influence of dehydrating agents on the oxidative carbonylation of methanol for dimethyl carbonate synthesis over a Cu/Y-zeolite catalyst. *Chin. J. Chem. Eng.* **2018**, *26*, 1059–1063. [CrossRef]
19. Li, Q.; Zhou, Z.; Chen, R.; Sun, B.; Qiao, L.; Yao, Y.; Wu, K. Insights into the reaction mechanism of CO oxidative coupling to dimethyl oxalate over palladium: A combined DFT and IR study. *Phys. Chem. Chem. Phys.* **2015**, *17*, 9126–9134. [CrossRef] [PubMed]
20. Song, Y.; Zhang, J.; Lv, J.; Zhao, Y.; Ma, X. Hydrogenation of Dimethyl Oxalate over Copper-Based Catalysts: Acid–Base Properties and Reaction Paths. *Ind. Eng. Chem. Res.* **2015**, *54*, 9699–9707. [CrossRef]
21. Zhu, Y.; Wang, S. Hydrogenation of dimethyl oxalate to ethylene glycol over Cu/SiO_2 catalysts. In Proceedings of the 2011 International Conference on Electronics, Communications and Control (ICECC), Ningbo, China, 9–11 September 2011; pp. 4344–4347. [CrossRef]
22. Niu, Y.; Li, C.; Shen, J.; Wei, X. Absorption of dilute sulfur dioxide in ethanediamine with ethylene glycol or polyethylene glycol 400 plus water system. *J. Clean. Prod.* **2018**, *171*, 506–511. [CrossRef]
23. Kim, S.H.; Hong, S.H. Transfer Hydrogenation of Organic Formates and Cyclic Carbonates: An Alternative Route to Methanol from Carbon Dioxide. *ACS Catal.* **2014**, *4*, 3630–3636. [CrossRef]
24. Lee, W.; Lee, J.W. Concurrent Production of Carbon Monoxide and Manganese(II) Oxide through the Reaction of Carbon Dioxide with Manganese. *ACS Sustain. Chem. Eng.* **2014**, *2*, 1503–1509. [CrossRef]
25. Gmehling, J.; Möllmann, C. Synthesis of Distillation Processes Using Thermodynamic Models and the Dortmund Data Bank. *Ind. Eng. Chem. Res.* **1998**, *37*, 3112–3123. [CrossRef]
26. Fredenslund, A.; Jones, R.L.; Prausnitz, J.M. Group-contribution estimation of activity coefficients in nonideal liquid mixtures. *AIChE J.* **1975**, *21*, 1086–1099. [CrossRef]
27. Shen, W.F.; Benyounes, H.; Song, J. A review of ternary azeotropic mixtures advanced separation strategies. *Theor. Found. Chem. Eng.* **2016**, *50*, 28–40. [CrossRef]
28. Fogler, H.S. *Essentials of Chemical Reaction Engineering*; Prentice-Hall: Upper Saddle River, NJ, USA, 2017.
29. Zabetakis, M.G.; Scott, G.S.; Jones, G.W. Limits of Flammability of Paraffin Hydrocarbons in Air. *Ind. Eng. Chem.* **1951**, *43*, 2120–2124. [CrossRef]
30. Chang, Y.-M.; Tseng, J.-M.; Shu, C.-M.; Hu, K.-H. Flammability studies of benzene and methanol with various vapor mixing ratios at 150 °C. *Korean J. Chem. Eng.* **2005**, *22*, 803–812. [CrossRef]
31. Carlson, E.C. Don't gamble with physical properties. *Chem. Eng. Prog.* **1996**, *92*, 35.
32. Wei, H.-M.; Wang, F.; Zhang, J.-L.; Liao, B.; Zhao, N.; Xiao, F.-K.; Wei, W.; Sun, Y.-H. Design and Control of Dimethyl Carbonate–Methanol Separation via Pressure-Swing Distillation. *Ind. Eng. Chem. Res.* **2013**, *52*, 11463–11478. [CrossRef]
33. Li, M.; Xu, X.; Li, X.; Ma, K.; Qin, B.; Zhu, Z.; Wang, Y. Phase Behavior and Thermodynamic Model Parameters in Simulations of Extractive Distillation for Azeotrope Separation. *Sci. Rep.* **2017**, *7*, 9497. [CrossRef] [PubMed]
34. Umbach, J.S.; Nitsche, L.S. Online Pinch Analysis Tool Pinch Analysis. Available online: http://www.uic-che.org/pinch/ (accessed on 12 December 2021).
35. Holtbruegge, J.; Kuhlmann, H.; Lutze, P. Process analysis and economic optimization of intensified process alternatives for simultaneous industrial scale production of dimethyl carbonate and propylene glycol. *Chem. Eng. Res. Des.* **2015**, *93*, 411–431. [CrossRef]
36. El-Halwagi, M.M. *Process Integration*, 1st ed.; Elsevier: Oxford, UK, 2006; Volume 7.
37. Olsen, D.; Abdelouadoud, Y.; Liem, P.; Wellig, B. The Role of Pinch Analysis for Industrial ORC Integration. *Energy Procedia* **2017**, *129*, 74–81. [CrossRef]
38. Ebrada, L.C.; De Luna, M.D.G.; Manegdeg, F.G.; Grisdanurak, N. The effect of minimum temperature difference in the design and optimization of heat exchanger networks of a brewery based on Pinch Methodology. *Lect. Notes Eng. Comput. Sci.* **2014**, *2*, 952–957.
39. Turton, R.; Bailie, R.C.; Whiting, W.B.; Shaeiwitz, J.A.; Bhattacharyya, D. *Analysis, Synthesis, and Design of Chemical Processes*; Pearson: Upper Saddle River, NJ, USA, 2012.
40. Jenkins, S. 2019 Chemical Engineering Plant Cost Index Annual Average. 2020. Available online: https://www.chemengonline.com/2020-annual-cepci-average-value/ (accessed on 20 February 2022).

processes

MDPI

Article

Biolubricant Production through Double Transesterification: Reactor Design for the Implementation of a Biorefinery Based on Rapeseed

José María Encinar, Sergio Nogales-Delgado * and Antonio Pinilla

Department of Chemical Engineering and Physical-Chemistry, University of Extremadura, Avda. De Elvas s/n, 06006 Badajoz, Spain; jencinar@unex.es (J.M.E.); apinillazy@alumnos.unex.es (A.P.)
* Correspondence: senogalesd@unex.es

Abstract: The production and use of biolubricants as replacements for mineral lubricants align with the promotion of sustainable development goals, contributing to the sustainable economic growth of developing countries as well as the preservation of the environment. The implementation of biorefineries (where the production of biolubricants through transesterification could play an important role) is becoming important for these purposes, using natural feedstocks such as rapeseed, which is an interesting crop that can adapt to different climates under extreme weather conditions. The aim of this research work was to study the design of a reactor to produce rapeseed biolubricant through double transesterification of the corresponding vegetable oil. Thus, the kinetics to design a reactor was studied, sizing it according to the properties of the biolubricant and the demand in Spain. As a result, a SAE 10W30 biolubricant was obtained, which was suitable for Diesel engines. A batch reactor was selected for the production of this biolubricant at a national level, with a pseudo-first reaction order and a reactor volume of 9.66 m^3.

Keywords: Brassica Napus; fatty acid methyl esters; 2-Ethyl-2-hydroxymethyl-1,3-propanediol; viscosity; batch reactor; kinetics

Citation: Encinar, J.M.; Nogales-Delgado, S.; Pinilla, A. Biolubricant Production through Double Transesterification: Reactor Design for the Implementation of a Biorefinery Based on Rapeseed. *Processes* **2021**, *9*, 1224. https://doi.org/10.3390/pr9071224

Academic Editor: Jean-Claude Assaf

Received: 31 May 2021
Accepted: 13 July 2021
Published: 15 July 2021

1. Introduction

A biolubricant is a lubricant that is neither detrimental to health nor harmful to the environment. Their main advantages, compared to the use of conventional lubricants, are their adaptation to green chemistry, biodegradability, sustainability and compliance with current legislation and customers' needs, focused on environmentally friendly products [1]. Apart from that, other physical properties that are improved compared to mineral lubricants are: better lubricity, higher flash and combustion points, or higher viscosity index, among others. However, there are some disadvantages such as their lower oxidative stability and cold flow properties [2]. Nevertheless, the use of additives like antioxidants can improve the performance of this product [3].

The main raw materials for biolubricant production are vegetable oils, such as rapeseed, sunflower, soybean, safflower, etc., which are mainly composed of triglycerides. In that sense, rapeseed is an interesting oil crop, as it is resistant to cold and hot climates in drylands, having short vegetation periods and a fatty acid composition that is suitable for food use, among others. Moreover, one interesting characteristic of this crop is its long tap root, which makes it suitable to take part in crop rotations [4]. Finally, its high oleic acid content (with one single double bond, which makes it less prone to oxidation compared to linoleic or linolenic acid) could contribute to the stability of the final products obtained through transesterification, such as biodiesel or biolubricants. Indeed, the influence of the raw material on these products has been widely studied in the literature, presenting different viscosity or oxidative stability values depending on the fatty acid profile [5,6].

There are several processes to obtain products (from vegetable oils) that can be used as biolubricants, such as selective hydrogenation, estolide formation, epoxidation or double transesterification [3,7]. Concerning epoxidation, consisting of the interconnection of unsaturated C–C bonds by an oxygen atom (at low reaction temperatures), the biolubricants obtained presented better lubricity and thermo-oxidative stability. However, pour point increased, whereas viscosity index was reduced. Regarding transesterification, it is the reaction of an ester and an alcohol to produce another ester through alkyl group exchange. Normally, the biolubricants obtained have good low temperature properties and high thermo-oxidative stability, whereas high reaction temperatures are required [3]. Figure 1 shows this process. As it can be seen, the first transesterification implies the use of triglycerides and methanol to produce fatty acid methyl esters (FAMEs, biodiesel) and glycerol. For the second transesterification, FAMEs react with a superior alcohol to produce the corresponding ester, releasing methanol.

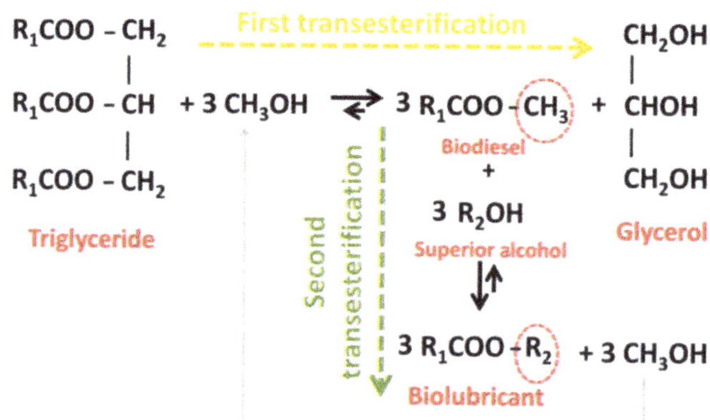

Figure 1. Double transesterification for biolubricant production. R_1 = alkyl group of fatty acids. R_2 = alkyl group of the superior alcohol.

As a consequence, double transesterification could provide plenty of products and by-products that can be valuable during biolubricant production. The whole process shows a high atom economy, as most by-products can be re-used in this same process or valued for market purposes, which is a considerable advantage. In that manner, there are many positive aspects in this process that can be suitable for the implementation of biorefineries [8]:

- Use of natural and renewable sources.
- Production of valuable products (like biodiesel and biolubricant).
- Generation of exploitable by-products (such as glycerol, with a wide range of uses depending on its purity [9]).
- Reutilization of by-products (like methanol).

This way, biorefineries can contribute to the economic growth of many developing countries or less-industrialized areas. In that sense, the Extremadura region is a traditionally agricultural area in Spain, which can profit from circular economy and current environmental policies, where the installation of biorefineries can perfectly fit, as pointed out in previous studies [10].

However, the implementation of biorefineries might be a challenge, as many factors have to be taken into account, such as raw materials, technologies, processing routes, products, and technical, economic or environmental aspects [8]. This way, the design of different components for biorefineries can help assess the feasibility of the implementation of these facilities. As far as we know, few specific studies have been carried out in order to prove the feasibility of biorefinery implementation based on vegetable oils through the design of its components (such as reactors), determining the kinetics and production for a certain area or region depending on the consumption of the products obtained.

The aim of this study was to assess the role of rapeseed oil as a starting point for a biorefinery, focusing on biolubricant production through double transesterification. Thus, the kinetics of the second transesterification as well as the reactor design and sizing (according to a specific production of the final biolubricant) were carried out.

2. Materials and Methods

2.1. Biodiesel and Biolubricant Production

Rapeseed oil was produced in CICYTEX (Centro de Investigaciones Científicas y Tecnológicas de Extremadura) by the Department of Energy and Extensive Crops. The oil had high-quality parameters such as low free fatty acid content (<1%), only requiring filtration after mechanical extraction. Afterwards, the oil was stored in 25-L opaque containers for further treatments and analysis.

Rapeseed biolubricant was obtained through double transesterification with different alcohols. Thus, the steps were the following (see Table 1 for the main chemical conditions):

- First transesterification with methanol. For fatty acid methyl ester production (biodiesel), the chemical conditions were based on previous studies [11].
- Second transesterification with 2-ethyl-2-hydroxymethyl-1,3-propanediol (Sigma-Aldrich, Saint Louis, MO, USA; solid state; molar mass, 134.18 $g \cdot mol^{-1}$; density, 1.08 $g \cdot cm^{-3}$; melting point, 56–58 °C; vapor pressure, 67 hPa at 200 °C; flash and combustion points, 180 and 367 °C, respectively). The biolubricant production was optimized in order to obtain high yields, exceeding 95%. For that purpose, experimental sets at different FAME/alcohol ratios (1:1/3; 1:1/2, 1:1 and 1:2), temperatures (100, 110 and 120 °C) and catalyst concentrations (1.0, 1.5 and 2.0% *w/w*) were carried out. These experiments were also used for the kinetic determination and reactor design. Once the reaction took place, the sample was purified by vacuum filtration, separating the surplus alcohol.

Table 1. Transesterification conditions for biolubricant production.

Reaction	Reaction Time, min	Reaction Temperature, °C	Alcohol/Oil Ratio [1]	Catalyst Concentration [2], %
First transesterification	150	60	6:1	1.5
Second transesterification	90	120	1:1	1.5

[1] The alcohol used for the first transesterification was methanol, whereas for the second transesterification 2-ethyl-2-hydroxymethyl-1,3-propanediol was used. [2] The catalyst used in both cases was sodium methylate.

2.2. Biodiesel and Biolubricant Characterization

Both biodiesel and biolubricant were characterized according to the UNE-EN 14,214 standard [12], as explained in previous studies [2,11,13]. Density was determined by using a pycnometer, correcting the value at 15 °C. Viscosity was measured by using Ostwald's viscometer at 40 and 100 °C, obtaining with these data the viscosity index. Flash and combustion points were obtained through the open-cup Cleveland method. Moisture was determined by using a Metrohm 870 KF Trinitro Plus (Metrohm, Herisau, Switzerland). Acid and iodine numbers were determined following the corresponding volumetric determination, according to the standard and cold filter plugging point (CFPP) was obtained (when necessary) for the samples obtained [13–15]. Regarding the yield of the reactions and fatty acid methyl ester profile of biodiesel, gas chromatography was used, following the conditions explained elsewhere [16]. Thus, the yield of the biolubricant was obtained by the decrease in FAMEs. Equally, the reaction evolution was also assessed by FAME decrease. Finally, oxidative stability was calculated through the Rancimat method, obtaining the induction point (expressed in hours) [17].

2.3. Kinetic Study and Volume Reactor Determination

In order to design a reactor for biolubricant production from biodiesel, different factors were taken into account, as follows:

- Characterization of the biolubricant, which determines its use in industry, paying special attention to viscosity at 40 and 100 °C and viscosity index and comparing with traditional lubricants.
- Determination of the order of reaction of the second transesterification process, obtaining the main kinetic parameters. For that purpose, the decrease in FAMEs from biodiesel to biolubricant production at different reaction times (0, 5, 15, 25 and 90 min) was measured to determine the reaction evolution. Different parameters like temperature and catalyst addition were studied.
- Calculation of biolubricant production at a national level (in this case, Spain), comparing with the equivalent lubricant production at the same level.
- Selection of the suitable reactor depending on production, obtaining the design equation by a mass balance.
- Reactor volume determination.

3. Results and Discussion

3.1. Biodiesel Characterization

First, a characterization of the biodiesel used as a starting point for biolubricant production was carried out. As it can be observed in Table 2, this biofuel complied with almost all the parameters established in the UNE-EN 14,214 standard, except for the oxidative stability. Thus, the chemical conditions were suitable to obtain a high-yield product, showing a high flash point, which implies safety during its storage.

Table 2. Main features of rapeseed biodiesel and comparison with the standard (green tick = compliance; wrong red cross = non-compliance).

Property	Units	Rapeseed Biodiesel	EN-14214 Limit Lower	EN-14214 Limit Upper	Compliance with the Standard
FAME content	%	98.4	96.5	-	✓
Moisture	mg/kg	250	-	500	✓
Density at 15 °C	kg/m^3	868	860	900	✓
Viscosity at 40 °C	mm^2/s	4.90	3.50	5.00	✓
Flash point	°C	194	101	-	✓
Oxidative stability	h	5.37	8	-	✗
Acid number	mg KOH/g	0.35	-	0.50	✓
Iodine number	g I_2/100g	105	-	120	✓
Methyl linolenate	% (m/m)	9.1	-	12	✓
CFPP	°C	−4	−10	0	✓

Regarding FAME content of biodiesel, the main components were methyl oleate (66.4%), linoleate (16.6%), linolenate (9.1%) and palmitate (3.8%). The higher content in methyl oleate, which is a mono-unsaturated compound, could explain the higher oxidative stability compared to other biodiesel samples found in the literature, needing lower amounts of antioxidant additives to comply with the standard [18–20].

3.2. Reaction Optimization

Concerning the optimization of the second transesterification reaction, Figures 2–4 show the different conversion evolution depending on the different parameter changes (FAME/2-ethyl-2-hydroxymethyl-1,3-propanediol ratio; temperature and catalyst concentration, respectively). In the case of FAME/alcohol ratio, Figure 2 shows the main results, obtaining similar yields for 1:1 and 1:2 ratios. As the amount of alcohol increased, the reaction rate increased for the initial stages. The 1:2 ratio was discarded, as it implied higher costs and operational difficulties (due to the filtration process, which was more difficult as the amount of surplus alcohol was higher), selecting 1:1 ratio for further studies, as in the case of previous studies with similar alcohols (2,2-dimethyl-1,3-propanediol) [10].

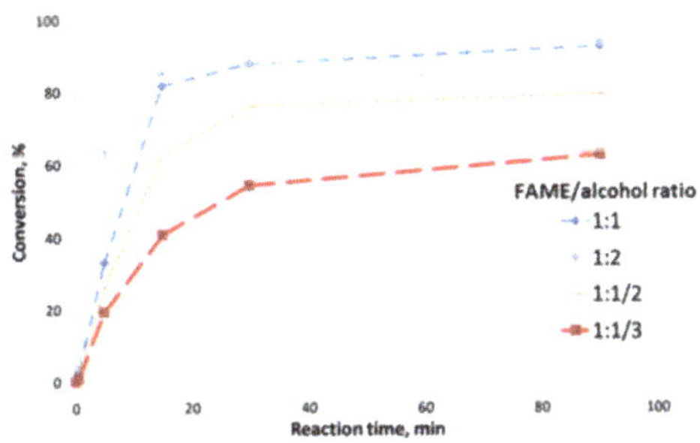

Figure 2. Effect of different FAME/alcohol ratios (1:1, 1:2; 1:1/2 and 1:1/3) on converScheme 100 °C; catalyst concentration, 1%; stirring rate, 500 rpm; reaction time, 90 min).

Regarding temperature, Figure 3 shows the differences between the reaction temperatures selected for this experience. Thus, as temperature increased, the reaction took place at a faster rate, achieving high conversions (exceeding 80%) in 15 min, reaching a balance at 25 min for all cases. This behavior was also observed for other biolubricants (obtained from rapeseed biodiesel and 2-ethyl-1-hexanol), where temperature considerably increased the reaction rate at initial stages [2]. For the final production of biolubricant, the temperature at which the highest yield was obtained (exceeding 94%) was selected, that is, 120 °C, which is a relatively low reaction temperature compared to other biolubricant production processes observed in the literature, reaching 150 °C [18].

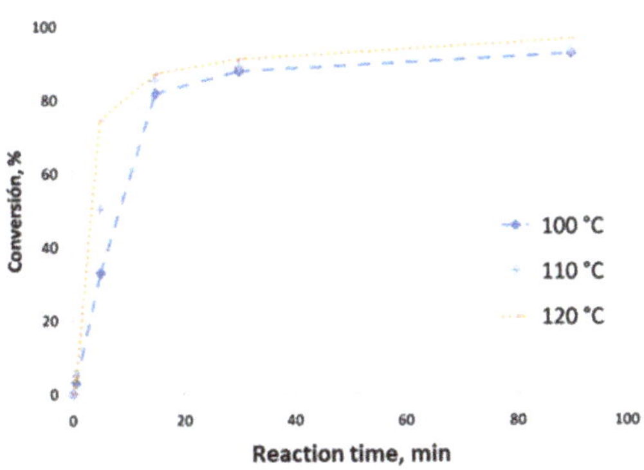

Figure 3. Effect of temperature. (FAME:alcohol ratio, 1:1; catalyst concentration, 1%; stirring rate, 500 rpm; reaction time, 90 min).

Finally, Figure 4 shows the effect of catalyst concentration. As expected, the rate of the reaction was higher as the catalyst addition was higher (and the equilibrium was achieved at shorter reaction times), reaching similar final conversions for 1.5 and 2.0% *w/w*. Consequently, an intermediate catalyst addition (1.5%) was selected, following economic criteria. Again, similar trends were obtained according to the literature, where the use of catalysts promoted the generation of products at faster reaction rates [2,10]. This way, it was proved that the use of catalysts is vital to make these processes competitive compared to traditional productions based on mineral products.

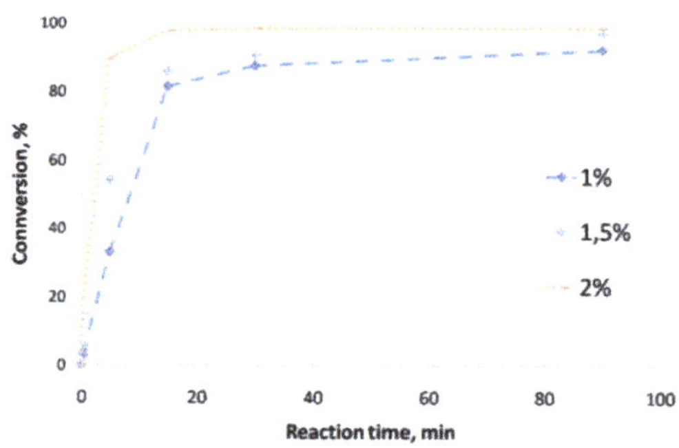

Figure 4. Effect of catalyst concentration (FAME:alcohol ratio, 1:1; reaction temperature, 100 °C; stirring rate, 500 rpm; reaction time, 90 min).

3.3. Biolubricant Characterization

In order to produce the final biolubricant, the best chemical conditions observed in the previous section (included in Table 1 for the second transesterification) were used, and the conversion evolution was included in Figure 5.

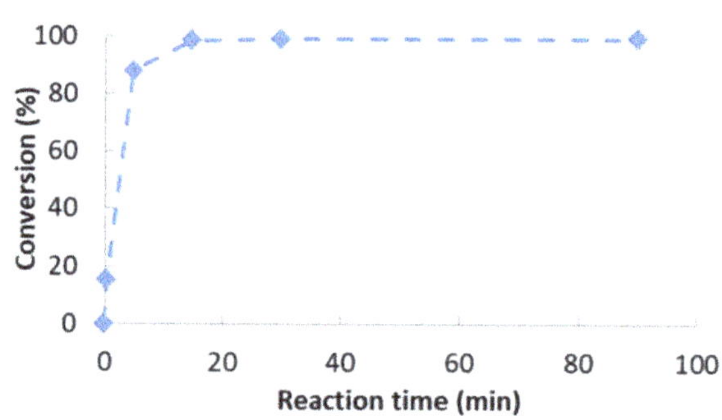

Figure 5. Conversion evolution during the second transesterification to obtain rapeseed biolubricant (FAME:alcohol ratio, 1:1; reaction temperature, 120 °C; catalyst concentration, 1.5%; stirring rate, 500 rpm; reaction time, 90 min).

Thus, conversion exceded 99%, with a considerable increase during the first 20 min of reaction (with around 97% at 15 min), obtaining a high-yield biolubricant under these chemical conditions. The main characteristics of the purified biolubricant are included in Table 3. According to the viscosity values obtained for 40 and 100 °C, this biolubricant could be comparable to a SAE10W30 lubricant (some characteristics of commercial lubricants are included in this table), which will be taken into account for the reactor design in following sections. Viscosity index, the resistance to viscosity changes over temperature, was relatively high compared to other products [21,22]. Flash and combustion points were high, within the range of other values found in the literature, which implies increased safety during storage [23,24]. According to data obtained for a SAE10W30 commercial lubricant (see Table 3), viscosity index and flash point were lower compared to these samples, although in the same order of magnitude.

Table 3. Main features of rapeseed biolubricant.

Feature [1]	Units	Rapeseed Biolubricant	Commercial Lubricant [1] (SAE 10W30)	Commercial Lubricant [2] (SAE 10W30)
Conversion	%	99.41	n.d. [2]	n.d.
Density at 15 °C	kg/m^3	949	859	869
Moisture	%	0.03	n.d.	n.d.
Acid number	mg KOH/g	0.28	n.d.	n.d.
Viscosity at 40 °C	mm^2/s	75.5	66	69.8
Viscosity at 100 °C	mm^2/s	10.7	10.4	10.5
Viscosity index	Dimensionless	128	145	137
Flash point	°C	210	232	228
Combustion point	°C	222	n.d.	n.d.
Oxidative stability	h	4.94	n.d.	n.d.

[1] All these determinations were carried out following the UNE-EN 14,214 standard. [2] n.d. = not determined.

3.4. Reactor Design

The choice of a continuous or discontinuous reactor mainly depends on the demand of the product obtained. In this case, a continuous reactor (steady state stirred tank reactor) was chosen, as it can be seen in Figure 6.

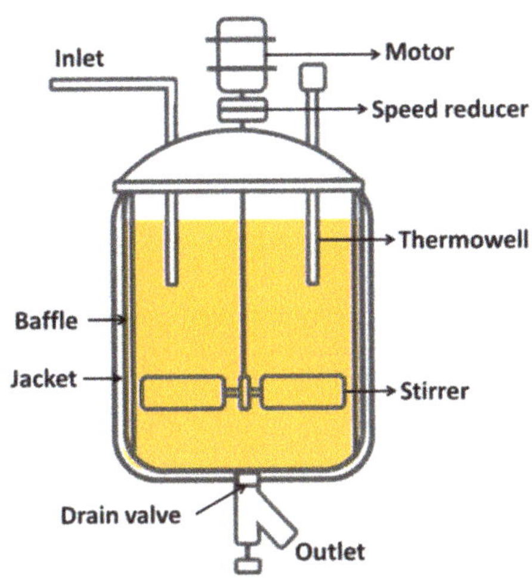

Figure 6. Main components of a continuous reactor.

The equation for reaction design is obtained through a mass balance. Figure 7 shows the basis for the design of a continuous reactor.

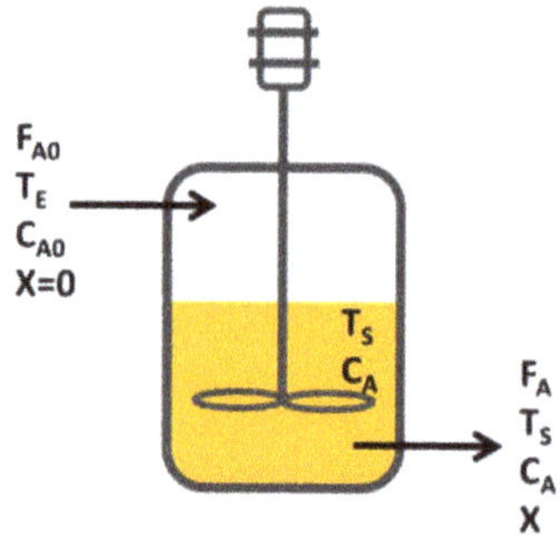

Figure 7. Scheme of a continuous reactor.

Where F_{A0} and F_A represent the mole flow of the limiting reagent at the reactor inlet and outlet, respectively; T_E and T_S are the temperatures of the stream at the reactor inlet and outlet, respectively; C_{A0} and C_A are the concentrations of the limiting reactive at the reactor inlet and outlet, respectively.

If the balance of the limiting reactive is done, Equation (1) is obtained:

$$F_{A0} = F_A + (-r_A)V_R + \frac{dN_A}{dt} \tag{1}$$

were $-r_A$ is the reaction rate and dN_A/dt is the accumulation of the limiting reagent. As the process is isothermal (there is not any density change) and it takes place in a steady state operation ($dN_A/dt = 0$), Equation (1) can be written in the form of Equation (2), where X_A represents the conversion of reactive A.

$$V_R = \frac{F_{A0} - F_A}{-r_A} = F_{A0} \cdot \frac{X_A}{-r_A} \tag{2}$$

In order to determine the volume of the reactor, it is necessary to know the flow of reactive A at the reactor inlet or outlet, the reaction kinetics and the final conversion. Regarding conversion, it was considered to be 99.41%, that is, $X_A = 0.9941$.

To determine F_{Ao}, some data about biolubricants were considered. According to ASELUBE, 111,000 tonnes of lubricants for automotive industry were sold [25]. On the other hand, according to other sources, the percentages of sales of SAE10W30 lubricants was 5% [26]. As a consequence, and considering that the obtained biolubricant is similar to SAE10W30 characteristics (according to data provided in Table 3), a production of 5550 tm/year was established, in order to cover the market demand in Spain.

To determine the mole flow of reagent A (biodiesel or fatty acid methyl ester), it is necessary to know the average molecular weight of biodiesel through biodiesel composition. Table 4 shows these data, obtaining an average molecular weight for rapeseed biodiesel of 284.29 g/mol.

Table 4. Average molecular weight of rapeseed biodiesel.

Methyl Ester	Percentage (%)	Molecular Weight (g/mol)	Weighted Molecular Weight (g/mol)
Oleate	66.4	296.49	196.87
Linoleate	16.6	294.48	48.88
Linolenate	9.10	292.50	26.62
Palmitate	3.80	270.46	10.28
Stearate	0.55	298.50	1.64
Total	96.45	-	284.29

Taking into account that the conversion of the process can be considered as 1 in practical terms, Table 5. shows the mass, volumetric and mole flows of biodiesel, 2-ethyl-2-hydroxymethyl-1,3-propanediol (alcohol) and sodium methoxide (catalyst). These data were obtained according to a yearly typical production of SAE10W30 lubricant in a chemical plant located in Extremadura region [25]. Accordingly, biodiesel, alcohol and catalyst flows were calculated according to the optimum reaction conditions found in this experience (Biodiesel/alcohol mole ratio, 1/1 and catalyst concentration, 1.5%).

Table 5. Flows at the inlet of the reactor.

Reactives	Mass Flow(tm/Year)	Volumetric Flow (m^3/Year)	Mole Flow (kmol/Year)
Biodiesel	5293.12	6098.07	18,618.75
Alcohol	2498.26	2313.21	18,618.75
Catalyst	116.87	120.49	2163.07
Total	7908.26	8531.76	39,400.56

3.5. Kinetic Study

The kinetic determination was carried out by applying the data analysis integral method. The reaction can be described according to Equation (3), where A is FAMEs, B is 2-ethyl-2-hydroxymethyl-1,3-propanediol, C is the biolubricant obtained and D is methanol.

$$A + \frac{1}{3}B \overset{\text{Catalyst}}{\rightarrow} \frac{1}{3}C + D \quad (3)$$

The reaction rate is given by Equation (4), where k is the kinetic constant, C_A is biodiesel concentration, C_B is alcohol concentration and α and β are the reaction orders for A and B.

$$-r_A = -\frac{dC_A}{dt} = k \cdot C_A^{\alpha} \cdot C_B^{\beta} \quad (4)$$

Taking into account the relationship between the reactant and its conversion and the stoichiometry of the reaction, θ_B can be defined as follows (Equations (5)–(7)):

$$C_A = C_{A0} \cdot (1 - X_A) \quad (5)$$

$$C_B = C_{A0} \cdot (\theta_B - \frac{1}{3}X_A) \quad (6)$$

$$\theta_B = \frac{C_{B0}}{C_{A0}} \quad (7)$$

where C_{A0} and C_{B0} are the initial concentration of A and B, and X_A is FAME conversion. Replacing Equations (5)–(7) in Equation (4), the following expression is obtained (Equation (8)).

$$\frac{dX_A}{dt} = k \cdot C_{A0}^{\alpha+\beta-1} \cdot (1 - X_A)^{\alpha} \cdot \left(\theta_B - \frac{1}{3}X_A\right)^{\beta} \quad (8)$$

This equation is differential, whose integration after establishing the following limits (t = 0, X_A = 0) and (t = t, X_A = X_A) was (Equation (9)):

$$\int_0^{X_A} \frac{dX_A}{(1 - X_A)^{\alpha} \cdot \left(\theta_B - \frac{1}{3}X_A\right)^{\beta}} = \int_0^t k \cdot C_{A0}^{\alpha+\beta-1} \cdot dt \quad (9)$$

The solution of this equation depends on α and β values. Although there are many possibilities, in the case of transesterification (as described in the literature) a first-order reaction regarding both reactants can be considered. Thus, α and β = 1. On the other hand, for most experiments, a θ_B value of 1 was selected, which implies that the alcohol concentration was three times higher than the stoichiometric value. Under these circumstances, the concentration of this reagent can be considered constant and the kinetic model, originally of 1 + 1 order, can be considered a pseudo-first order reaction, that is, a model where the starting kinetic equation is shown in Equation (10), where k* is a complex constant that includes catalyst and alcohol concentration effects.

$$-r_A = -\frac{dC_A}{dt} = k^* \cdot C_A \tag{10}$$

Equation (10) can be expressed as a function of conversion, obtaining Equation (11):

$$\frac{dX_A}{dt} = k^*(1 - X_A) \tag{11}$$

As a result of the integration of Equation (11), Equation (12) is finally obtained, which is a linear equation, so the slope of the line is k*.

$$\ln\left[\frac{1}{1 - X_A}\right] = k^* \cdot t \tag{12}$$

Equation (12) has been tested with the experimental results obtained. For instance, Figure 8 shows the representation for one of the experiments carried out. Thus, the linear trend was observed for the first stages of the reaction, as the transesterification balance was achieved quickly and, under these circumstances, the net reaction rate (that is, the difference between the direct and inverse reaction rate) is close to zero. As a consequence, the reaction model represented in Equation (3), which assumes an irreversible reaction, does not make any sense at longer times, where the reversible reaction could take place.

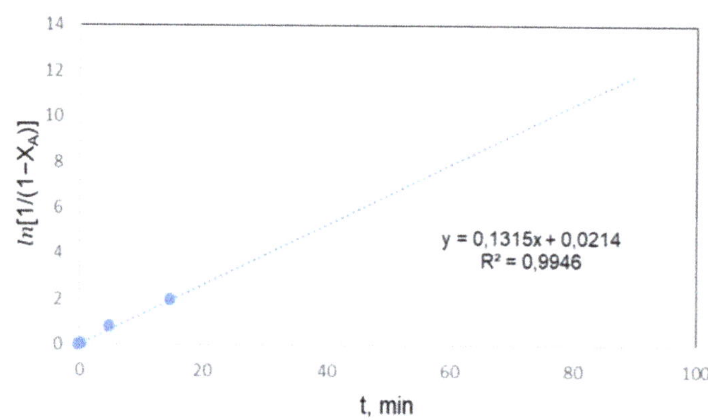

Figure 8. Linear representation of Equation (12) for the following chemical conditions: **FAME/alcohol** mole ratio, 1:1; catalyst concentration, 1.5% *w/w*; temperature, 100 °C; reaction time, 90 min; stirring rate, 500 rpm.

Thus, through least squares adjustment, Table 6 shows the pseudo-kinetic constant, which was obtained from the slope of the linear representation observed in Figure 8, and the coefficient of determination, corresponding to experiments with different catalyst concentration.

Table 6. Adjustment of data depending on catalyst concentration. Pseudo-first reaction order.

Units	1%	1.5%	2%
k^*, min^{-1}	0.1213	0.1315	0.2512
R^2	0.9869	0.9946	0.9242

As it can be observed, the pseudo-first reaction order fits reasonably, as long as the reaction time considered is short (for the first 15–20 min).

As it was pointed out, k* is a complex constant, including catalyst and alcohol concentrations. If the order of reaction of alcohol and catalyst is assumed to be the unit, k* would be expressed by Equation (13), and according to this equation, a representation of k* versus catalyst concentration should imply a line passing though the origin.

This representation is included in Figure 9, which shows a linear trend (with a low R^2 value), passing through the origin as the kinetic model predicted.

$$k^* = k \cdot C_{cat} \cdot C_B \tag{13}$$

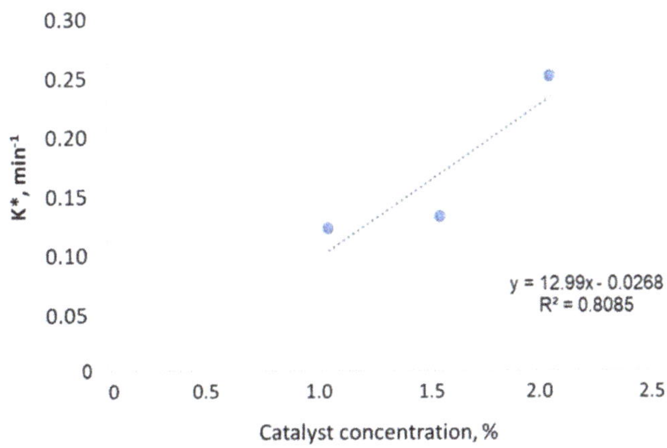

Figure 9. Dependence of k* with catalyst concentration (pseudo-first reaction order).

An important aspect in every kinetic study is the dependence of kinetic constants on temperature. To determine this dependence, experiments at 100, 110 and 120 °C were carried out, adjusting the evolution of the reaction to a pseudo-first reaction order. Thus, Table 7 shows the kinetic constants (k*) and R^2 coefficients corresponding to this experimental set. The adjustment is reasonable, especially for the two first experiments, whereas in the case of the experiment carried out at the highest temperature the adjustment was worse, due to the fact that the balance was achieved earlier (at shorter reaction times) and, therefore, fewer experimental values were suitable to carry out the adjustment. Thus, points at longer reaction times (30–90 min) cannot be considered as the inverse reaction started to be considerable (which is not assumed in the kinetic model included in this study).

Table 7. Adjustment data for different reaction temperatures. Pseudo-first reaction order.

Units	100 °C	110 °C	120 °C
k^*, min^{-1}	0.1213	0.1269	0.1359
R^2	0.9869	0.9984	0.8819

Although k* is a complex constant, it is possible to assess its dependence on temperature. Thus, considering Equation (13), k constant would be the real kinetic constant. Assuming that k presents a dependence on temperature given by the Arrhenius equation, Equation (14) can be obtained from Equation (13), where all the terms that remain constant with temperature were regrouped, obtaining (Cte).

$$k^* = k \cdot C_{cat} \cdot C_B = k_o . e^{-\frac{E}{RT}} . C_{cat} \cdot C_B = (Cte) e^{-\frac{E}{RT}} \tag{14}$$

Calculating the Napierian logarithm, Equation (15) is obtained:

$$\ln k^{*} = -\frac{E}{R}\cdot\frac{1}{T} + \ln(Cte) \quad (15)$$

where E is the activation energy, R is the ideal gas constant, T is absolute temperature and (Cte) is a constant where all the constant terms are included, including the pre-exponential factor k_0.

From Equation (15), through a least square adjustment of data corresponding to the experiments carried out at different temperatures, it is possible to determine the activation energy of the process. Table 8 shows the adjustment data, the activation energy and R^2 value. It should be pointed out the low activation energy, which is in accordance with the fact that it is a catalytic reaction carried out at relatively high temperatures (100–120 °C). This low activation energy value implies a very low dependence of the reaction rate on temperature. Thus, in the temperature range that has been studied, the reaction rate is not very sensitive to temperature. In any case, and even though according to previous results an optimization of reaction temperature would be justified, the reactor design, as explained in the following section, was carried out for a reaction temperature of 120 °C.

Table 8. Arrhenius equation. Adjustment data at 100, 110 and 120 °C by using Equation (15).

Slope (−E/R)	Intercept	Activation Energy	R^2
−831.7	0.1151	6.91 kJ/mol	0.9821

Finally, considering the kinetic data obtained, the theoretical conversion corresponding to the final chemical conditions was determined. Thus, in Figure 10, a comparison between the experimental and theoretical data is included, pointing out a good convergence and ratifying the reliability of the kinetic data obtained. As it can be seen, the adjustment is especially good at the beginning of the reaction, for the first 10 min, where high conversions were observed. As a consequence, the kinetic model proposed could be suitable for the subsequent reactor design, as explained in further sections.

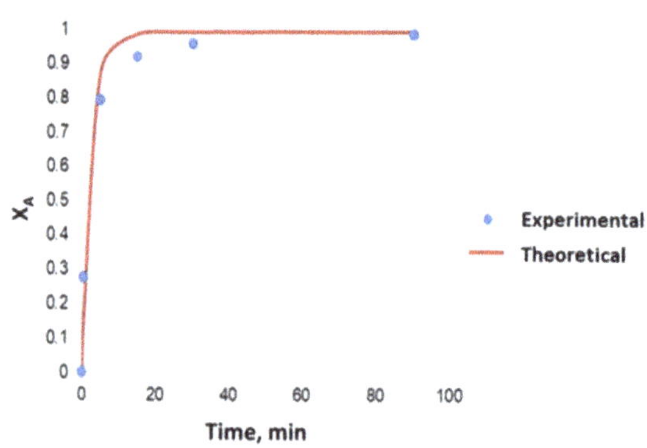

Figure 10. Theoretical and experimental conversions for the final experiment (FAME/alcohol mole ratio, 1/1; catalyst concentration, 1.5%; reaction temperature, 120 °C; reaction time, 90 min; stirring rate, 500 rpm).

Thus, the reactor design will be carried out with the abovementioned chemical conditions corresponding to the final biolubricant production, that is, 1:1 mole ratio, 1.5% catalyst concentration, 120 °C, a reaction time of 90 min and a stirring rate of 500 rpm.

3.6. Reactor Volume

Reactor volume was determined by Equation (2). In order to apply this equation, F_{A0}, X_A and $-r_A$ should be known. F_{A0} was previously determined (18,618.75 kmol/year). If expressed in international units, 0.5904 mol/s are obtained. For X_A, the conversion obtained in the final experiment was considered (0.9941). In order to calculate $-r_A$, Equation (10) was used, which implies the calculation of C_A and k*. C_A value, that is, the concentration of fatty acid methyl esters at the reactor outlet, can be obtained from Equation (5), previously obtaining C_{A0} with total volumetric (and mole) flow (included in Table 5). This way, C_{A0} = 2182.3 mol/m^3 and C_A = 12.9 mol/m^3. Finally, k* was obtained by using Equation (12) as previously explained. Table 9 shows the calculations made.

Table 9. Reactor volume determination.

Variable	Variable Calculation, Units
Inlet mole flow, F_{Ao}	F_{Ao} = **18,618.75** kmol·year^{-1}= **0.5904 mol·s^{-1}**
Conversion, X_A	X_A = 0.9941 (obtained from final experiment)
Inlet concentration, C_{Ao}	$C_{Ao} = F_{Ao}/F_{Vtotal}$ = 2182.3 mol/m^3
Outlet concentration, C_A	$C_A = C_{Ao}(1- X_A)$ = 12.9 mol/m^3
Pseudo-first reaction order constant, k*	k* = 0.2827 min^{-1} = 4.7117.10^{-3}.s^{-1}
Reaction rate, $-r_A$	$-r_A$ = k* C_A = 4.7117·10^{-3}·12.9 = 6.078·10^{-2} mol/m^3.s
Reactor volume, V_R	$V_R = F_{Ao}·X_A/-r_A$ = 0.5904·0.9941/6.078·10^{-2} = 9.6564 m^3

According to the reactor volume, a 20% oversizing of the equipment was considered. Therefore, the reactor should have a volume of 12 m^3.

4. Conclusions

The main findings were the following:

- Rapeseed biodiesel complied with almost all the standard requirements, highlighting the high flash and combustion points (exceeding 190 °C). However, oxidative stability did not comply with the standard (5.37 h), although it was relatively high compared to other equivalent samples found in the literature.
- The optimum chemical conditions to obtain high yields of rapeseed biolubricant were biodiesel/alcohol (2-ethyl-2-hydroxymethyl-1,3-propanediol) mole ratio: 1/1; catalyst (sodium methoxide) concentration, 1.5% *w/w*; reaction temperature, 120 °C; reaction time: 90 min. Under these circumstances, high yields (exceeding 99%) were obtained.
- The characteristics of the biolubricant were calculated, being equivalent to a SAE 10W30 lubricant, suitable for Diesel engines.
- For the reactor design, due to the operating conditions, the kind of reaction and desired production, a continuous reactor was selected.
- The kinetics of the process could be considered a pseudo-first order reaction, obtaining an activation energy of 6.91 kJ·mol^{-1}.
- According to the annual production (once the kinetics was known), a reactor volume of 9.66 m^3 was obtained (12 m^3 if oversizing is considered).

Author Contributions: Conceptualization, J.M.E. and S.N.-D.; methodology, J.M.E. and S.N.-D.; validation, J.M.E. and S.N.-D.; formal analysis, S.N.-D. and A.P.; investigation, S.N.-D. and A.P.; resources, J.M.E.; data curation, J.M.E., S.N.-D. and A.P.; writing—original draft preparation, S.N.-D. and A.P.; writing—review and editing, J.M.E. and S.N.-D.; visualization, J.M.E.; supervision, J.M.E. and S.N.-D.; project administration, J.M.E.; funding acquisition, J.M.E. All authors have read and agreed to the published version of the manuscript.

Funding: This research was funded by the "Junta de Extremadura" and "FEDER" (IB18028 and GR18150).

Institutional Review Board Statement: Not applicable.

Informed Consent Statement: Not applicable.

Acknowledgments: The authors would like to thank the "Junta de Extremadura" and "FEDER" for the financial support received to perform this study.

Conflicts of Interest: The authors declare no conflict of interest.

Nomenclature

C_{Ao}	Inlet concentration
C_A	Outlet concentration
F_{Ao}	Inlet mole flow
F_A	Outlet mole flow
k^*	Pseudo-first reaction order constant
$-r_A$	Reaction rate
T_E	Inlet temperature
T_s	Outlet temperature
V_R	Reactor volume
X_A	Conversion

References

1. Salimon, J.; Salih, N.; Yousif, E. Industrial development and applications of plant oils and their biobased oleochemicals. *Arab. J. Chem.* **2012**, *5*, 135–145. [CrossRef]
2. Encinar, J.M.; Nogales-Delgado, S.; Sánchez, N.; González, J.F. Biolubricants from rapeseed and castor oil transesterification by using titanium isopropoxide as a catalyst: Production and characterization. *Catalysts* **2020**, *10*, 366. [CrossRef]
3. Cecilia, J.A.; Plata, D.B.; Saboya, R.M.A.; de Luna, F.M.T.; Cavalcante, C.L.; Rodríguez-Castellón, E. An overview of the biolubricant production process: Challenges and future perspectives. *Processes* **2020**, *8*, 257. [CrossRef]
4. Beyzi, E.; Gunes, A.; Buyukkilic Beyzi, S.; Konca, Y. Changes in fatty acid and mineral composition of rapeseed (*Brassica napus* ssp. oleifera L.) oil with seed sizes. *Ind. Crop. Prod.* **2019**, *129*, 10–14. [CrossRef]
5. Chozhavendhan, S.; Vijay Pradhap Singh, M.; Fransila, B.; Praveen Kumar, R.; Karthiga Devi, G. A review on influencing parameters of biodiesel production and purification processes. *Curr. Res. Green Sustain. Chem.* **2020**, *1–2*, 1–6. [CrossRef]
6. Martínez, G.; Sánchez, N.; Encinar, J.M.; González, J.F. Fuel properties of biodiesel from vegetable oils and oil mixtures. Influence of methyl esters distribution. *Biomass Bioenergy* **2014**, *63*, 22–32. [CrossRef]
7. McNutt, J.; He, Q.S. Development of biolubricants from vegetable oils via chemical modification. *J. Ind. Eng. Chem.* **2016**, *36*, 1–12. [CrossRef]
8. Moncada, B.J.; Aristizábal, M.V.; Cardona, A.C.A. Design strategies for sustainable biorefineries. *Biochem. Eng. J.* **2016**, *116*, 122–134. [CrossRef]
9. Checa, M.; Nogales-Delgado, S.; Montes, V.; Encinar, J.M. Recent advances in glycerol catalytic valorization: A review. *Catalysts* **2020**, *10*, 1279. [CrossRef]
10. Nogales-Delgado, S.; Sánchez, N.; Encinar, J.M. Valorization of Cynara Cardunculus L. Oil as the Basis of a Biorefinery for Biodiesel and Biolubricant Production. *Energies* **2020**, *13*, 5085. [CrossRef]
11. Nogales-Delgado, S.; Encinar, J.M.; Guiberteau, A.; Márquez, S. The Effect of Antioxidants on Corn and Sunflower Biodiesel Properties under Extreme Oxidation Conditions. *J. Am. Oil Chem. Soc.* **2019**, *97*, 201–212. [CrossRef]
12. ISO. *UNE-EN 14214:2013 V2+A1:2018 Liquid Petroleum Products—Fatty acid Methyl Esters (FAME) for Use in Diesel Engines and Heating Applications—Requirements and Test Methods*; International Organization of Standardization: Geneva, Switzerland, 2018.
13. ISO. *UNE-EN-12634:1999 Productos Petrolíferos y Lubricantes. Determinación del Índice de Ácido. Método de Valoración Potenciométrica en un Medio no Acuoso*; International Organization of Standardization: Geneva, Switzerland, 1999.
14. ISO. *UNE-EN 14111:2003 Fat and Oil Derivatives. Fatty Acid Methyl Esters (FAME)*; Determination of Iodine Value; International Organization of Standardization: Geneva, Switzerland, 2003.
15. ISO. *UNE-EN 116:2015 Diesel and Domestic Heating Fuels—Determination of Cold Filter Plugging Point—Stepwise Cooling Bath Method*; International Organization of Standardization: Geneva, Switzerland, 2015.
16. ISO. *UNE-EN ISO 12966-2:2011 Animal and Vegetable Fats and Oils—Gas Chromatography of Fatty Acid Methyl Esters—Part 2: Preparation of Methyl Esters of Fatty Acids*; International Organization of Standardization: Geneva, Switzerland, 2011.
17. Focke, W.W.; Van Der Westhuizen, I.; Oosthuysen, X. Biodiesel oxidative stability from Rancimat data. *Thermochim. Acta* **2016**, *633*, 116–121. [CrossRef]
18. Encinar, J.M.; Nogales, S.; González, J.F. Biodiesel and biolubricant production from different vegetable oils through transesterification. *Eng. Rep.* **2020**, 1–10. [CrossRef]
19. Varatharajan, K.; Pushparani, D.S. Screening of antioxidant additives for biodiesel fuels. *Renew. Sustain. Energy Rev.* **2018**, *82*, 2017–2028. [CrossRef]
20. Dunn, R.O. Antioxidants for improving storage stability of biodiesel. *Biofuels Bioprod. Biorefin.* **2008**. [CrossRef]

21. Da Silva, J.A.C.; Soares, V.F.; Fernandez-Lafuente, R.; Habert, A.C.; Freire, D.M.G. Enzymatic production and characterization of potential biolubricants from castor bean biodiesel. *J. Mol. Catal. B Enzym.* **2015**, *122*, 323–329. [CrossRef]
22. Encinar, J.M.; Nogales, S.; González, J.F. Biorefinery based on different vegetable oils: Characterization of biodiesel and biolubricants. In Proceedings of the 3rd International Conference in Engineering Applications (ICEA), Ponta Delgada, Portugal, 8–11 July 2019.
23. Salimon, J.; Salih, N.; Yousif, E. Biolubricant basestocks from chemically modified ricinoleic acid. *J. King Saud Univ. Sci.* **2012**, *24*, 11–17. [CrossRef]
24. Kania, D.; Yunus, R.; Omar, R.; Abdul Rashid, S.; Mohamad Jan, B. A review of biolubricants in drilling fluids: Recent research, performance, and applications. *J. Pet. Sci. Eng.* **2015**, *135*, 177–184. [CrossRef]
25. ASELUBE Asociación Española de Lubricantes. Available online: http://aselube.net (accessed on 31 May 2020).
26. Statista.com. Lubricant Sales in Spain. Available online: https://www-statista-com.ezproxy.unex.es (accessed on 31 May 2021).

processes

Article

Investigation of the Jet Characteristics and Pulse Mechanism of Self-Excited Oscillating Pulsed Jet Nozzle

Si Zhang, Biwei Fu * and Lin Sun

School of Mechanical Engineering, Yangtze University, Jingzhou 434023, China; zhangsi@yangtzeu.edu.cn (S.Z.); 202072577@yangtzeu.edu.cn (L.S.)
* Correspondence: fubiwei@yangtzeu.edu.cn

Abstract: Self-excited oscillation pulse jet technology is widely used to clean sediment from oil storage tanks. Its successful application is dependent on jet performance. As the cleaning requirements of the oil industry increase, it is necessary to optimise the structure of self-excited oscillation pulsed jet nozzles (SOPJNs) to optimise cleaning and energy efficiencies. In this study, the jet performance of a SOPJN is modelled and analysed based on computational fluid dynamics with consideration of a large eddy simulation and homogeneous cavitation. The modelling results are highly consistent with experimental results. The effects of the SOPJN's inlet diameter, cavity diameter, cavity length, wall reflection angle, and inlet pressure on the jet's peak velocity, oscillation frequency, and cavitation number were analysed. The results show that the oscillation frequency decreases with the increase of the inlet diameter d_1, cavity diameter D, cavity length L and reflection angle of wall α. Optimisation of the SOPJN inlet diameter, cavity length, and wall reflection angle produced a jet with a high peak velocity and strong cavitation. The optimal nozzle cavity diameter strengthens cavitation, while the peak velocity fluctuates as the cavity diameter increases. The peak velocity increases with the inlet pressure, while the increasing rate of the peak velocity decreases. The results of this study can be used in the design and optimisation of similar nozzle structures for improved pulse jet cleaning.

Keywords: self-excited oscillation pulse jet; nozzle; large eddy simulation; peak velocity; cavitation number

Citation: Zhang, S.; Fu, B.; Sun, L. Investigation of the Jet Characteristics and Pulse Mechanism of Self-Excited Oscillating Pulsed Jet Nozzle. *Processes* **2021**, *9*, 1423. https://doi.org/10.3390/pr9081423

Academic Editor: Jean-Claude Assaf

Received: 17 June 2021
Accepted: 14 August 2021
Published: 17 August 2021

1. Introduction

Crude oil storage is an essential part of oil and gas processing. Long-term storage of crude oil in a tank will result in the formation of sediment at the bottom. If not promptly removed, this sediment can seriously affect the quality of the oil and accelerate corrosion of the inner wall of the tank floor, which could potentially lead to an oil spill, fire, or other severe accident. Therefore, tank-cleaning technology is key to safe oil and gas storage [1,2]. The main tank-cleaning methods include heating of the whole tank, chemical cleaning, heat oil circulation, and crude oil injection. Problems associated with these methods include long work periods, high energy consumption and environmental pollution [3–6]. Compared with the above methods, self-excited oscillation pulsed jet cleaning technology has many advantages, such as being low-cost, energy-efficient, safe, and environmentally friendly. A current topic of interest is the improvement of self-excited oscillation pulse jet cleaning technology, which can provide lower cost and higher efficiency than existing methods. The self-excited oscillation pulsed jet nozzle (SOPJN) is the core component of such systems. The structure of the nozzle directly affects the cleaning performance and cost, so it is important to analyse the effects of nozzle geometry and dynamics on jet performance.

There have been many studies on SOPJNs, which have mainly focused on three topics. (1) The mechanism of self-excited oscillation pulsed jets has been researched. (2) The influences of various factors on jet velocity and oscillation frequency have been analysed. (3) The influences of different factors on the cavitation capability of the nozzle have been analysed. For example, Thomas et al. [7–9] used theoretical analysis and numerical simulation to explain the mechanism of self-excited oscillation pulsed jets. By combining large eddy simulation with experiments, Fang et al. [10] studied the three-dimensional cavitation turbulence flow inside a Helmholtz oscillator. The results show that there is a close relationship between the vortex-cavitation interaction and the flow mechanism. Liu et al. [11] designed a self-excited aspiration pulsed jet device based on a traditional low-frequency self-excited oscillation pulsed jet device and revealed its impacts, characteristics, and pulse mechanism through experiments. Wang et al. [12–14] analysed the effects of the structural parameters of nozzles on the jet's peak velocity and oscillation frequency through numerical simulations and experiments and obtained optimised SOPJN structural parameters. Tamaki et al. [15–17] determined the main influences on an SOPJN's cavitation capability through numerical simulations and experiments. Xiang et al. [18] researched the periodical dynamics induced by the geometric parameters and operational variables of a self-excited pulsed cavitation jet under optimum experimental conditions. They revealed the mechanism of cavitation jet occurrence by considering the jet shape, striking force, and cavity pressure. Huang and Li et al. [19,20] analysed the influences of oscillating cavity parameters on the rock-breaking performance of an oscillating pulsed jet through experiments and simulations. Together, these studies found that the jet performance of a nozzle can be evaluated by examining the peak velocity, oscillation frequency, and cavitation capability. A limitation of existing research is that only one or two indicators of jet performance have been analysed. Most researchers have only studied the relationships between nozzle structural parameters and SOPJN performance qualitatively and not quantitatively. Accordingly, the present study aims to obtain better jet performance from nozzles by comprehensively and quantitatively analysing the influences of their structural parameters on jet performance.

The CFD method is used widely in the research on cavitating jets. At present, there are three main numerical simulation models of such turbulence: Direct numerical simulations (DNS), Reynolds-averaged Navier–Stokes (RANS) models and large eddy simulations (LES). A DNS can obtain accurate information of the turbulent field and is an effective way to study turbulence mechanisms. However, existing computational resources are often insufficient for high-Reynolds-number flow simulations, which limits their application. The RANS models can calculate complex flows with high Reynolds numbers but cannot reflect the detailed information of the flow field. The LESs are based on the mechanism of turbulent kinetic energy transmission. They can directly calculate the motion of large-scale vortices and describe the influence of small-scale vortices on large-scale vortices via modelling. They not only obtain more dynamic information than RANS models, such as the structure and properties of large-scale vortices but also have higher computational efficiency than DNS. It is considered to be one of the best potential turbulent flow numerical simulation methods. Many experts have used LES and the homogeneous flow model to predict turbulent flow cavitation. The results of LES are highly consistent with those of experiments [21–28]. Thus, we chose to use the LES approach in our research. In Section 2, we will explain the main factors affecting the performance of self-excited oscillation jets in detail. We then describe the CFD method, explain the control equation, and present a two-dimensional calculation model based on fluid properties and boundary conditions. Lastly, we experimentally and theoretically verify the results of the numerical model. In Section 3, we compare the impacts of different nozzle structural parameters on the jet's peak velocity, oscillation frequency, and cavitation number. And we examine the jet performance of the optimised nozzle under different operating parameters. In Section 4, we discuss the main conclusions of this investigation.

2. Numerical Models and Evaluation Indexes of Jet Performance

2.1. Numerical Analysis Model

2.1.1. Governing Equations

Numerical simulation is used in this paper as it can clearly monitor the internal and external flow field characteristics of the nozzle. In addition, the effects of different conditions and different nozzle structure parameters on the jet flow field can be analysed and discussed in an all-round way. Compared with the experiments, it greatly reduces the research cost and time. Therefore, the numerical simulation is also applied in this research. To clearly understand the mechanism of a self-excited oscillation pulse jet and the characteristics of cavitation bubbles and hydrodynamics, an LES model was used to analyse the flow field of the nozzle with consideration of the transition between gas and liquid.

1. LES control equations

In LES equations, flow field variables are divided into large-scale pulses and small-scale pulses. To obtain control equations for large-scale pulses, filtering is adopted to eliminate small-scale pulses from the turbulent flow. Then, the additional stress of the sub-grid scale is introduced into the control equation to solve the Navier–Stokes equation. Finally, the small-scale pulses are solved by additional equations. The large-scale pulses and small-scale pulses are respectively defined as the solvability scale and grid scale of the turbulence flow.

(1) Navier–Stokes equations

Navier–Stokes equations are used to describe the pulse jet. After dimensionless processing, the fluid density ρ is equal to 1 and the reciprocal of the Reynolds number is the viscosity coefficient ν. The altered Navier–Stokes equations are shown in Equation (1).

$$\begin{cases} \frac{\partial u_i}{\partial t} + u_j \frac{\partial u_i}{\partial x_j} = \nu \frac{\partial^2 u_i}{\partial x_j \partial x_j} - \frac{1}{\rho}\frac{\partial p}{\partial x_i} - F \\ \frac{\partial u_i}{\partial x_i} = 0 \end{cases} \tag{1}$$

(2) Filter functions

Under the LES framework, a physical quantity f can be divided into two parts in the two-dimensional incompressible turbulent flow field. The component $\overline{f}$ is the large-scale pulse and component f' is the small-scale pulse. The relationship between f, $\overline{f}$ and f' is described by Equation (2). The expression of $\overline{f}$ is found in Equation (3). The Gaussian filter function $G_i(x_i,x_i')$ is in the x_i-direction and its expression is shown as Equation (4). The grid length Δ_i is in the x_i-direction.

$$f = \overline{f} + f' \tag{2}$$

$$\overline{f}(x_1, x_2) = \iint_S \prod_{i=1}^{2} G_i(x_i, x_i') f(x_1', x_2') dx_1' dx_2' \tag{3}$$

$$G_i(x_i, x_i') = \left(\frac{6}{\pi \Delta_i^2}\right)^{1/2} \exp\left\{-6\frac{(x_i - x_i')^2}{\Delta_i^2}\right\} (i = 1, 2) \tag{4}$$

(3) LES control equations

After dimensionless processing, the Navier–Stokes equations are filtered to obtain the LES turbulence Equation (5). The sub-grid scale stress τ_{ij} is the filtered momentum transport that is generated by the small- and large-scale pulses.

$$\begin{cases} \frac{\partial \overline{u}_i}{\partial t} + \overline{u}_j \frac{\partial \overline{u}_i}{\partial x_j} = \frac{1}{\mathrm{Re}} \cdot \frac{\partial}{\partial x_j} \frac{\partial \overline{u}_i}{\partial x_j} - \frac{\partial \tau_{ij}}{\partial x_j} - \frac{\partial \overline{p}}{\partial x_i} \\ \frac{\partial \overline{u}_i}{\partial x_j} = 0 \end{cases} \tag{5}$$

(4) The sub-grid scale model

The Smagorinsky eddy viscosity model is adopted as the sub-grid scale model in this paper and is expressed as Equation (6). The eddy viscosity coefficient v_T can be solved by Equation (7).

$$\tau_{ij} = 2v_T\overline{S_{ij}} = v_T\left(\frac{\partial \overline{u}_i}{\partial x_j} + \frac{\partial \overline{u}_j}{\partial x_i}\right) \tag{6}$$

$$v_T = (C_s\Delta)^2\left(\frac{1}{2}\left(\frac{\partial \overline{u}_i}{\partial x_j} + \frac{\partial \overline{u}_j}{\partial x_i}\right)^2\right)^{1/2} \tag{7}$$

2. Homogeneous Equilibrium Model

A homogeneous equilibrium model (HEM) is used to analyse cavitation flow. In the HEM, a gas-liquid two-phase fluid can be considered to be a mixed homogeneous and variable-density single-phase fluid. β_l is the volume fraction of the liquid and β_v is the volume fraction of liquid vapour. For a control volume β_l, the value of β_v can range between 0 and 1, with their sum being 1. There is no cavitation phenomenon when $\beta_l = 1$ and there is complete cavitation when $\beta_l = 0$. ρ is the mixture's density and μ is the mixture's dynamic viscosity.

$$\rho = \beta_l\rho_l + \beta_v\rho_v \tag{8}$$

$$\mu = \beta_l\mu_l + \beta_v\mu_v \tag{9}$$

Under the precondition of isothermal homogeneous flow, the compressibility of the liquid and vapour are low in the cavitation flow, and they can be considered to be an incompressible fluid [29–31]. Based on the above assumptions, the control equations of the two-phase turbulent flow in the nozzle are as follows:

(1) Continuity equation:

$$\frac{\partial \rho}{\partial t} + \nabla\left(\rho\vec{u}\right) = 0 \tag{10}$$

(2) Momentum equation:

$$\frac{\partial(\rho\vec{u})}{\partial t} + \nabla\left(\rho\vec{u}\right) = -\nabla p + \nabla\left[\mu\left(\nabla\vec{u} + \nabla\vec{u}^{\Gamma}\right)\right] + \rho g + F \tag{11}$$

(3) Volume fraction equation:

$$\frac{\partial \beta_v}{\partial t} + \nabla\left(\beta_v\vec{u}\right) = \frac{\beta_l\rho_l}{\rho}\frac{4\pi n r^2}{1 + 4\pi n r^3/3}\frac{dr}{dt} \tag{12}$$

(4) Rayleigh established the Rayleigh–Plesset Equation (13) on the condition of incompressible flow. The equation can be used to calculate the process of bubble movement, growth, and collapse.

$$r\frac{d^2r}{dt^2}+\frac{3}{2}(\frac{dr}{dt})^2=\frac{p_{vap}-p}{\rho}-\frac{2\sigma}{\rho r}-\frac{4\mu}{\rho r}\cdot\frac{dr}{dt} \tag{13}$$

2.1.2. Calculation Model

1. Nozzle geometry and calculation setup

Figure 1 shows a SOPJN consisting of three main parts: the inlet nozzle, self-excited oscillation cavity (SOC), and outlet nozzle. The SOC is marked in blue with a red boundary in Figure 1. When a high-pressure water jet is injected into the SOC at high speed, the jet produces an unstable shear layer in the SOC, and many vortex rings are generated around the shear layer [32]. In this paper, the flow characteristics in the SOC of the self-excited oscillation pulsed jet nozzle are explored.

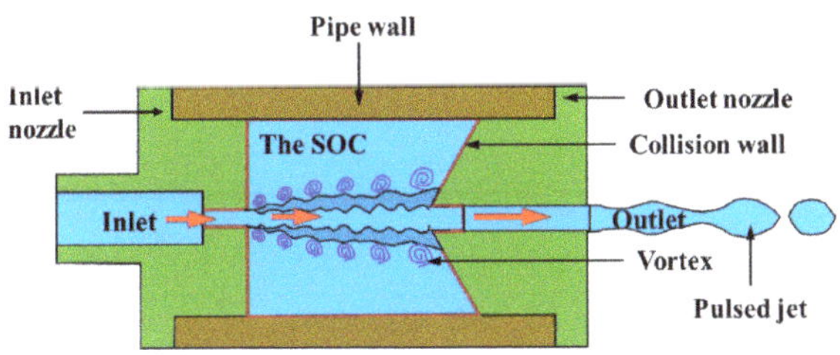

Figure 1. Schematic diagram of the self-excited oscillation pulsed jet nozzle.

The main characteristic parameters of the SOC are described in Figure 2a. They are the inlet diameter d_1, cavity diameter D, cavity length L, wall reflection angle α, and outlet diameter d_2. Considering the nozzle model to be axisymmetric, a two-dimensional computational model is used, as shown in Figure 2b. To reduce calculation errors caused by the outlet boundary condition, a large external flow field is established in the computational model. As shown in Figure 2b, the area surrounded by the red line is the extended external flow field. The boundary conditions and computational grids are described in detail in Figure 2b. The fluid domain is made up of 225,097 cells. A domain mesh refinement scheme is used in the central region of the calculation model and the minimum size is 0.2 mm.

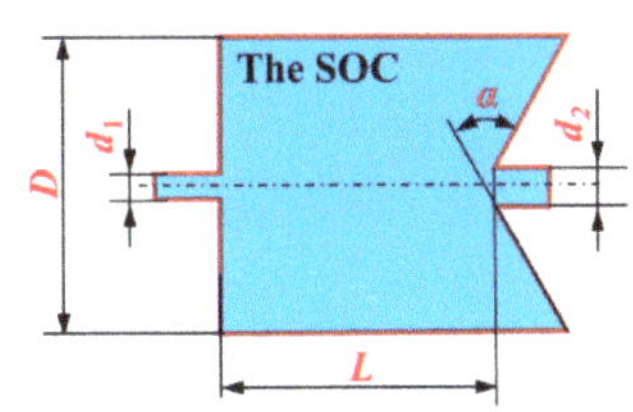

(**a**)

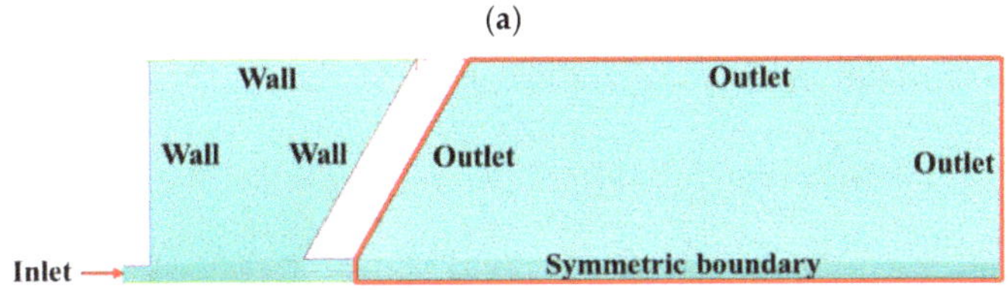

(**b**)

Figure 2. (**a**) Geometry and (**b**) gridding model and boundary conditions of the self-excited oscillation cavity (SOC).

2. Boundary conditions and fluid properties

It is well-known that establishing appropriate boundary conditions is important in ensuring the convergence and accuracy of simulations. In our case, the boundary conditions for the computation are shown in Figure 2b. Pressure boundary conditions are applied at the inlet and outlet edges. The boundary conditions of the walls are defined as being impermeable and not allowing slip. Symmetry conditions are employed at the symmetric boundary. The inlet pressure is set to range between 0 MPa and 3 MPa and atmospheric pressure is used at the pressure outlet.

The fluid in the simulations and experiments is water. The saturation pressure is set to 3540 Pa, the surface tension coefficient is 0.0717 N/m, and the vapour density is 0.02558 kg/m^3.

3. Calculation Settings

A monitoring point is selected on the centre axis of the calculation model, which is 50 mm apart from the exit of the outlet nozzle. The monitoring parameters include velocity, pressure, and vapour volume fraction. Based on the monitoring results, the cavitation numbers can be calculated by Equation (2).

2.2. Computational Model Validation

2.2.1. Experimental Verification

A principle diagram of the self-excited oscillation pulsed jet experiment is shown in Figure 3, in which (a) is a schematic diagram of the experimental system and (b) and (c) are the enlarged view of the region outlined by the red dotted line. The experimental purpose was to obtain the pressure and force at the monitoring point. The pressure and the force were measured by the same experimental system, but the measuring sensors were different. The pressure was measured by the pressure sensor, while the force was measured by the load cell. When the pressure was measured, the pressure sensor was installed in the experimental system, or the load cell was installed in the experimental system. The monitoring point of the maximum pressure and force are located at the centre of the end cover, which is 50 mm from the exit of the outlet nozzle, as respectively indicated in Figure 3b,c. In this experiment, the inlet diameter, cavity diameter, cavity length, angle of reflection of the wall, and outlet diameter of the nozzle are 8 mm, 100 mm, 35 mm, 60° and 11 mm, respectively. The experimental system is shown in Figure 4, which included a plunger pump, control cabinet, throttle valve, pressure gauge, pressure sensor, water tank, steel plate, and the SOPJN. The experiment measured the pressure and force at the monitoring point at inlet pressures of 1 MPa, 1.5 MPa, 2 MPa, 2.5 MPa and 3 MPa. The inlet pressure for each group was tested three times and the average value taken as the result.

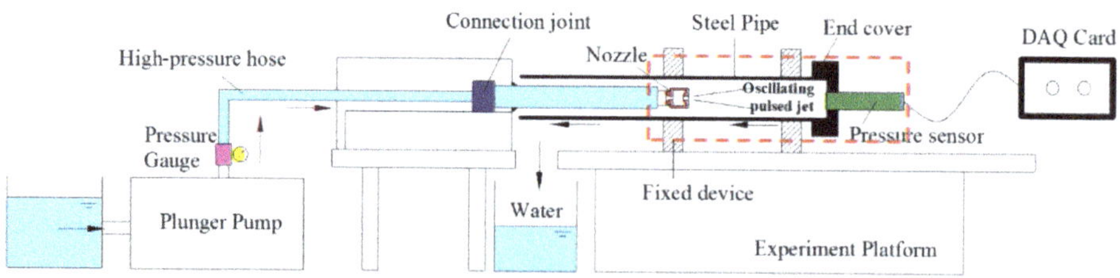

(a)

Figure 3. *Cont.*

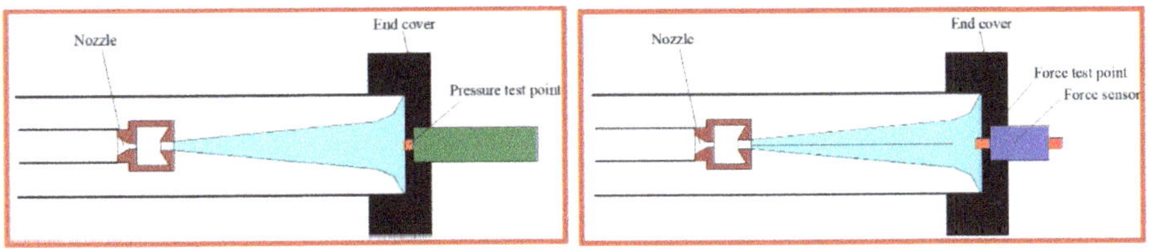

(**b**) (**c**)

Figure 3. (**a**) Schematic diagram of the experimental system and (**b**) enlargement of the red box containing the maximum pressure measurement point (**c**) enlargement of the red box containing the force measurement point.

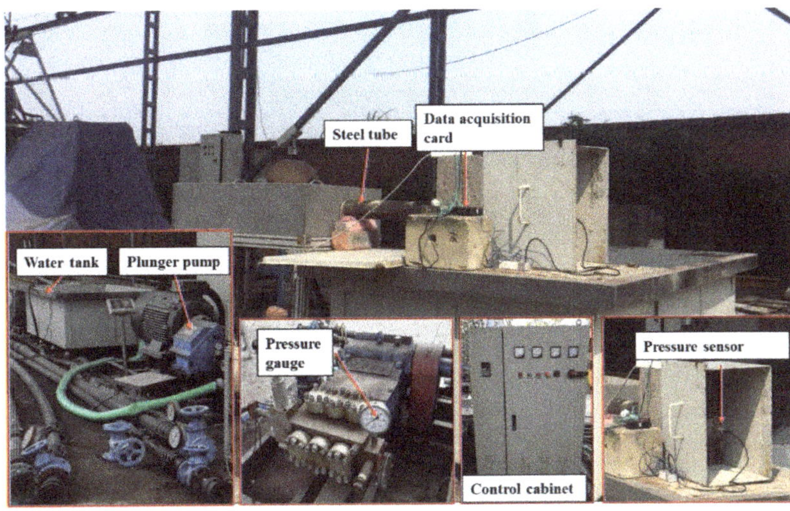

Figure 4. High-pressure water jet measurement system.

Figure 5 compares experimental and calculated results to verify the calculation model. Figure 5a,b show that the maximum pressure errors between the two results are <8.8% and the force errors are <6.6%, indicating high consistency. This demonstrates that the numerical calculation model can be used to analyse the jet performance of the nozzle.

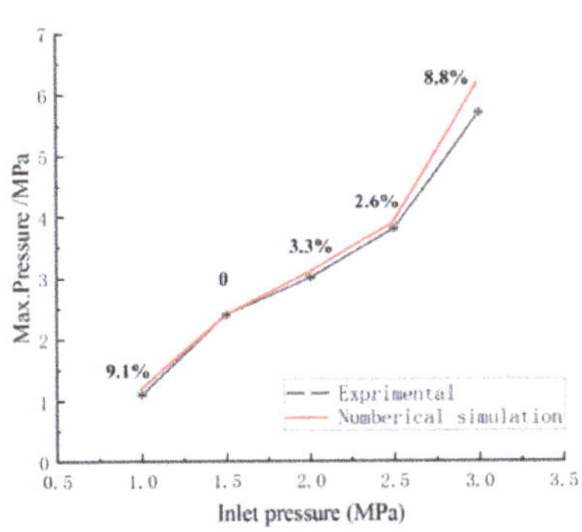

(**a**)

Figure 5. *Cont.*

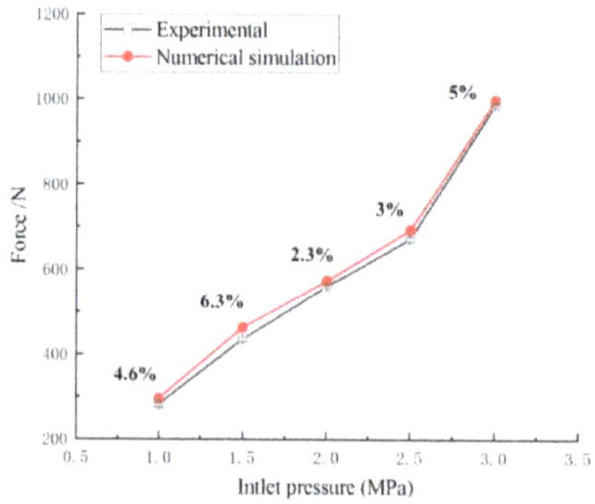

(b)

Figure 5. Errors between the experimental and calculated values of (**a**) maximum pressure and (**b**) force.

2.2.2. Mechanism Validation

According to boundary layer and vortex theory, vortex disturbance waves are generated in the separation zone when the jet passes through the SOC. The centre pressure of the vortex disturbance wave is low. For the self-excited oscillation pulsed jet to be generated, certain conditions must be satisfied within the oscillation cavity [24]. As shown in Figure 6, there is a process of change in pressure in the SOC. As shown in Figure 6a, the vortex disturbance wave is generated at the exit of the inlet nozzle, which is then amplified as it transmits downward. Next, the vortex disturbance wave impinges upon the collision wall and is reflected effectively. Finally, new vortex disturbance waves are generated at the exit of the inlet nozzle. These five conditions are shown in Figure 6 and demonstrate that the results of the computational model are robust.

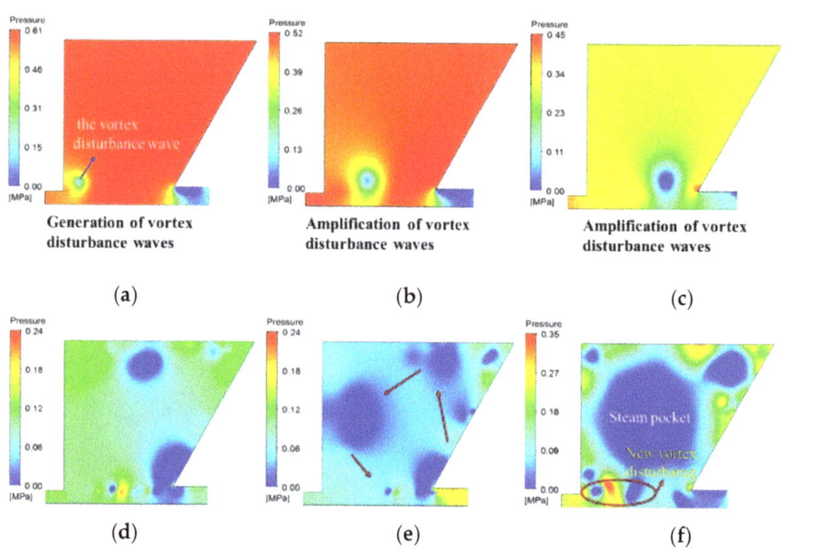

Figure 6. Pressure contours in the SOC at different times. (**a**) Generation of vortex disturbance waves (**b**) Amplification of vortex disturbance waves (**c**) Amplification of vortex disturbance waves (**d**) Interaction between wall and disturbance waves (**e**) Reflection of disturbance waves (**f**) Generation of new vortex disturbance waves and steam pocket.

Figure 7 clearly shows that disturbance waves, steam pockets, vortex rings, and pulses are generated in the SOC. The calculation results are consistent with the conclusions of previous studies [25,26]. The mechanism of the self-excited oscillation pulsed jet is shown in Figure 7. The vortex disturbance waves are generated by differences between the jet velocity and the SOC's inner fluid velocity. Some of the disturbance waves are amplified by the downward flow in the unstable shear layer. Next, the disturbance waves impinge upon the collision wall, then some are reflected effectively while others flow downwards with the jet. The reflected waves interact with the jet, which causes the impedance of fluid to cause periodic changes in the SOC; thus, pressure pulses occur in the flow. This process forms the self-excited oscillation pulsed jet.

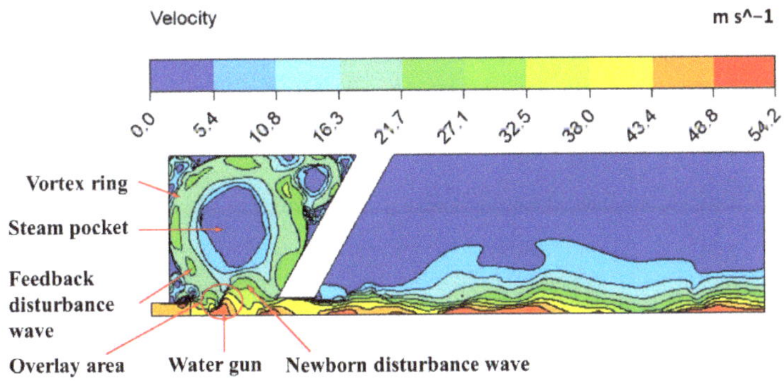

Figure 7. Velocity contours in the SOC.

Figure 8 shows the volume fraction of vapour in the computational model. It shows that a steam pocket and a series of bubbles are generated in the nozzle. According to vortex theory, bubbles are generated in the shear layer when the centre vortex pressure is lower than the saturated vapour pressure. Meanwhile, the number of bubbles is proportional to the number of vortices, because each vortex is characterized by a single cavity. Therefore, a greater number of bubbles corresponds to a larger number of vortices and a higher oscillation frequency [12,33].

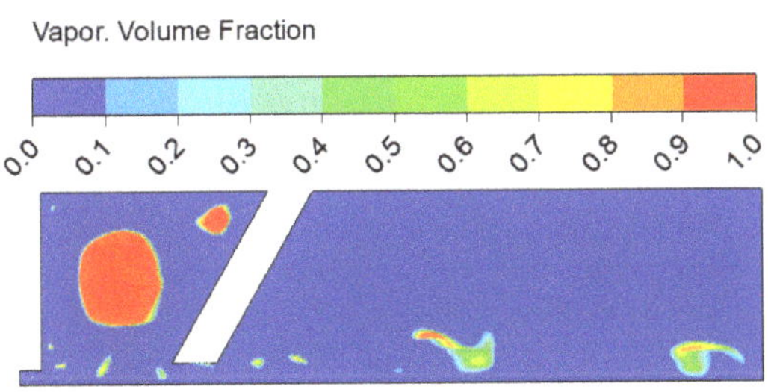

Figure 8. Vapour volume fractions in the SOC.

A temporal behavior of velocity at the monitoring point is researched. Figure 9 is the curve of velocity over time. As described in Figure 9, the axial velocity at the monitoring point oscillates around 40 m/s over time. This is an obvious pulsed jet characteristic.

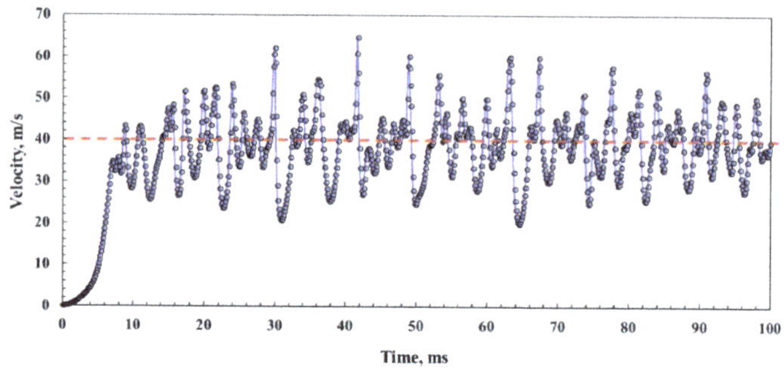

Figure 9. Change curve of axial velocity over time at the monitoring point.

2.3. *Jet Performance Evaluation Indexes*

The main evaluation indices of jet performance are peak velocity, cavitation number K and oscillation frequency ω.

1. Peak velocity

Jet velocity primarily consists of axial velocity and feedback velocity. Feedback velocity is a variable velocity induced by feedback disturbance waves. This paper assumes that the axial velocity has a base value and that the feedback velocity is the range of velocity amplitude. Fluctuation in the flow impedance appears in the SOC when the feedback disturbance waves interact with the jet and transforms it into a pulse jet. Higher feedback velocity results in a larger amplitude range and a greater oscillation effect. Peak velocity mainly affects the decontamination ability, with a higher peak velocity resulting in a greater decontamination effect.

2. Oscillation frequency

The expression of oscillation frequency is shown in Equation (14) [12], which shows that the oscillation frequency decreases with increases in inlet diameter, cavity diameter, and cavity length.

$$\omega = \frac{ad_0}{2\pi} \cdot \frac{\sqrt{1 + 1.63(d_2/d_1)^2}}{D\sqrt{L}} \tag{14}$$

3. Cavitation number

When the local pressure drops below the saturation pressure of the fluid, cavitation occurs, where bubbles go through a process of formation, development, and collapsing inside the fluid or at the liquid-solid interface. The cavitation number K is a dimensionless parameter that can be used to describe the cavitation state and features. It can also be used to determine the intensity of cavitation and to judge whether cavitation happens in the fluid or at the liquid-solid interface. The cavitation number is usually defined by Equation (15) and is a very large number at the initial stage of cavitation [34]. As cavitation develops further, the cavitation number gradually decreases. The critical cavitation number is defined as K_0. Cavitation will not occur unless K is less than K_0. A larger value of K_0 translates to a stronger capacity for cavitation.

$$K = \frac{p_i - p_s}{\rho u_i^2/2} \tag{15}$$

3. Results and Discussion

3.1. *Influences of Structural Parameters on Jet Performance*

3.1.1. Influence of Inlet Diameter on Jet Performance

1. Influence of inlet diameter on peak velocity and oscillation frequency.

With all other factors held constant in the computational model, the structure of the nozzle can be defined as d_1 (Inlet diameter, mm), D (Cavity diameter, mm), α (Reflection angle of wall, °), L (Cavity length, mm), d_2 (Outlet diameter, mm). In the analysis of influences of the inlet diameter d_1 on the jet performance, only inlet diameter d_1 is changing while the other structure parameters D, α, L, d_2 are kept constant. Here, the cases are respectively analysed when d_1 = 6 mm, 7 mm, 8 mm, 9 mm, 10 mm and 11 mm and the other structure parameters D = 100 mm, α = 60°, L = 35 mm, d_2 = 11 mm. The cases are uniformly denoted as d_1-100-60-35-11 (where d_1 = 6 mm, 7 mm, 8 mm, 9 mm, 10 mm and 11 mm respectively, and D = 100 mm, α = 60°, L = 35 mm, d_2 = 11 mm).

Figure 10 shows the change curves of peak velocity and feedback velocity with the inlet diameter d_1, while the cavity diameter d_2 is constant. The inlet flow rate and the velocity difference between the inlet jet and the fluid in the self-excited oscillation cavity are the two main factors affecting the formation and development of the disturbance wave. The larger the velocity difference is, the easier the disturbance wave will be formed. However, if the inlet flow rate is small and there is no sufficient energy, the disturbance wave is easy to collapse. Therefore, with increases in d_1/d_2 from 0.545 to 0.818, the inlet flow increase. It is beneficial to the development of the disturbance wave and then the feedback velocity increase at this stage. However, when the inlet diameter d_1 further increases, the velocity difference between the inlet jet and the fluid in the self-excited oscillation cavity is small. It results in the difficulty of the formation of disturbance waves. In addition, then the feedback velocity gradually decreases. On one hand, with increases in d_1/d_2, the feedback velocity first increases and then decreases. On the other hand, based on the fluid similarity network theory and the research of Li, the self-excited oscillation system has obvious resonance characteristics [32]. Therefore, with increases in d_1/d_2, the peak velocity has a similar sinusoidal variation. As shown in Figure 10, when d_1/d_2 is 0.545 and 0.818, the peak velocity is large. However, when d_1/d_2 is 0.545, the feedback velocity is very small, and the jet pulse effect is poor. Therefore, the optimal value of d_1/d_2 is 0.818. In this situation, both the peak velocity and feedback velocity reach their peak values and the oscillation effect is optimal.

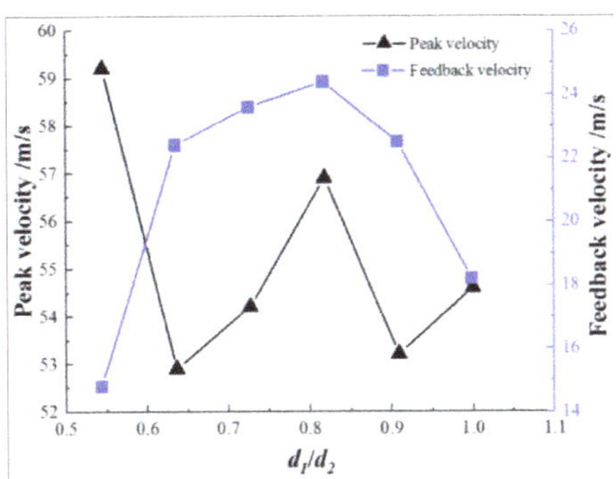

Figure 10. Curves of peak velocity and feedback velocity vs inlet diameter.

Figure 11 shows the influence of inlet diameter d_1 on the steam pocket region. It shows that as the inlet diameter d_1 increases, the serial bubbles and steam pocket region decrease. The initial results show that fewer serial bubbles result in a lower oscillation frequency. Therefore, with the increase of inlet diameter d_1, the oscillation frequency of the self-excited oscillation pulse jet nozzle decreases gradually.

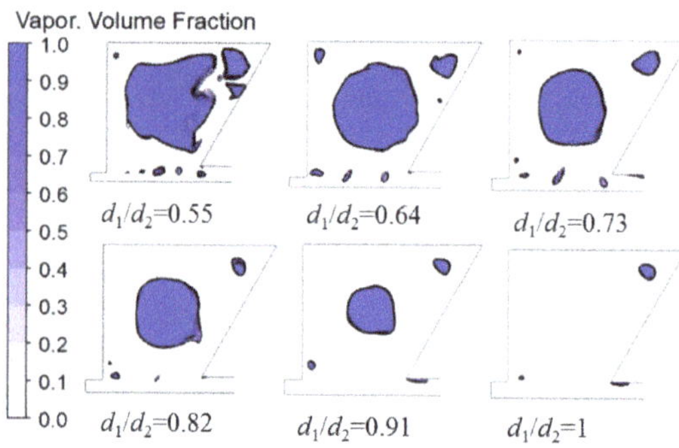

Figure 11. Steam pocket regions of the nozzle with different inlet diameters.

2. Influence of inlet diameter on cavitation number

To intuitively deduce the cavitation at the monitoring point, this paper assumes that when the vapour content of the monitoring point is greater than 0, the volume fraction of vapour will be set to 1; otherwise, the volume fraction of vapour is set to 0. According to this rule, the curves of the vapour volume fraction and cavitation number vary with time as shown in Figure 12. This leads to the conclusion that the value of K_0 is the maximum cavitation number. A greater value of K_0 means the system will be prone to easier cavitation. Figure 12 expresses that the critical cavitation number peaks at 1.42 when the inlet diameter is 9 mm, during which time the cavitation capability of the nozzle is the strongest.

3.1.2. Influence of Cavity Diameter on Jet Performance

1. Influence of cavity diameter on peak velocity and oscillation frequency

The influences of cavity diameter D on the jet performance were analysed using cases with dimensions 9-D-60-35-11 (where D = 85 mm, 90 mm, 100 mm, 105 mm and 110 mm respectively, and d_1 = 9 mm, α = 60°, L = 35 mm, d_2 = 11 mm). The curves of peak velocity and feedback velocity vs cavity diameter are shown in Figure 13. The figure shows that as the cavity diameter D increases, the feedback velocity first increases and then decreases. It can be explained according to the changes of steam pocket shown in Figure 14.

Figure 14 shows that the steam pocket region is affected by the cavity diameter D. When D/d_2 increases from 7.73 to 9.09, the steam pocket and serial bubbles both decrease. In addition, more energy is converted into the kinetic energy of the disturbance wave, then the feedback velocity gradually increases. When D further increases, a larger cavity diameter results in more fluid being contained within the cavity. If the inlet velocity and flow is the same, the total energy is the same. Then more fluid is contained in the steam pocket, which reduces the feedback velocity. Therefore, as the cavity diameter D increases, the feedback velocity increases first and then decreases. Feedback velocity and peak velocity peaks when D/d_2 is approximately 9 and the oscillation effect is the greatest.

2. Influence of cavity diameter on cavitation number

Figure 15 shows curves of the vapour volume fraction and cavitation number vs time. The graphs indicate that as the cavity diameter increases, K_0 initially increases and then decreases. K_0 peaks at 1.42 with a cavity diameter of 100 mm.

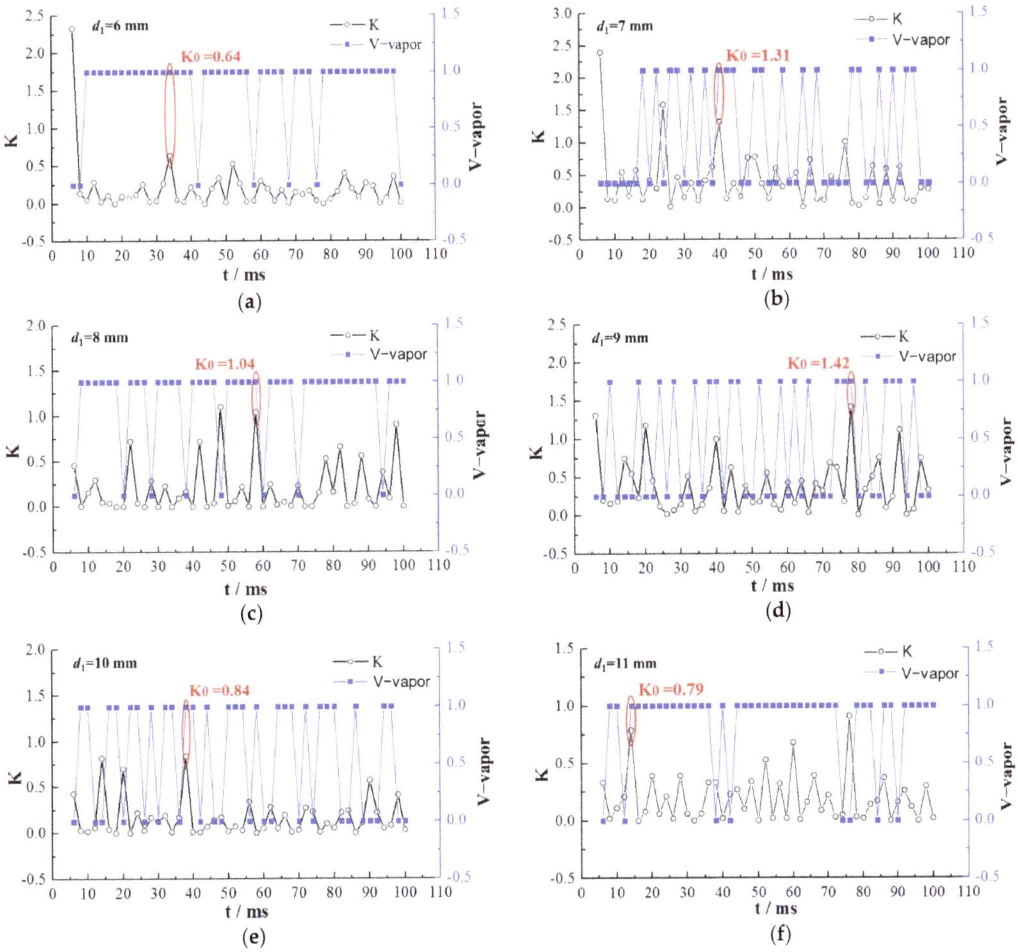

Figure 12. Cavitation number vs time for nozzles of various inlet diameters. (**a**) d_1 = 6 mm (**b**) d_1 = 7 mm (**c**) d = 8 mm (**d**) d_1 = 9 mm (**e**) d_1 = 10 mm and (**f**) d_1 = 11 mm.

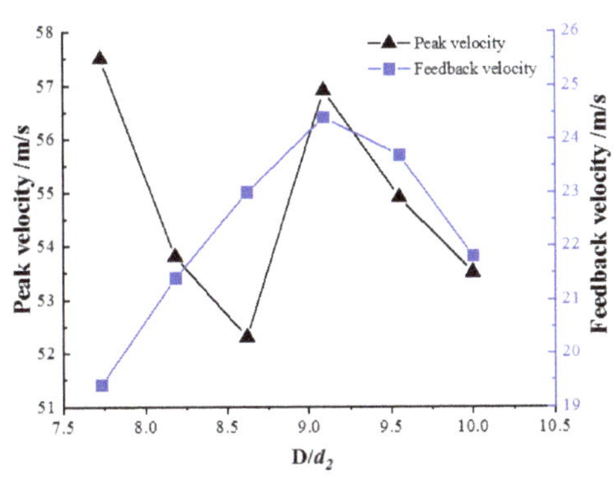

Figure 13. Curves of peak velocity and feedback velocity vs cavity diameter.

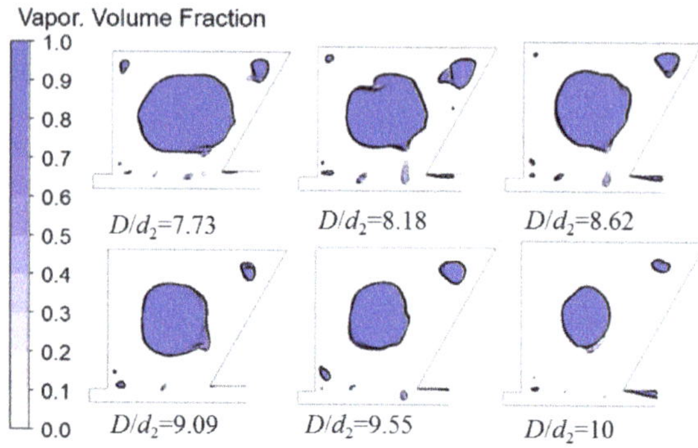

Figure 14. Steam pocket regions of the nozzle with different cavity diameters.

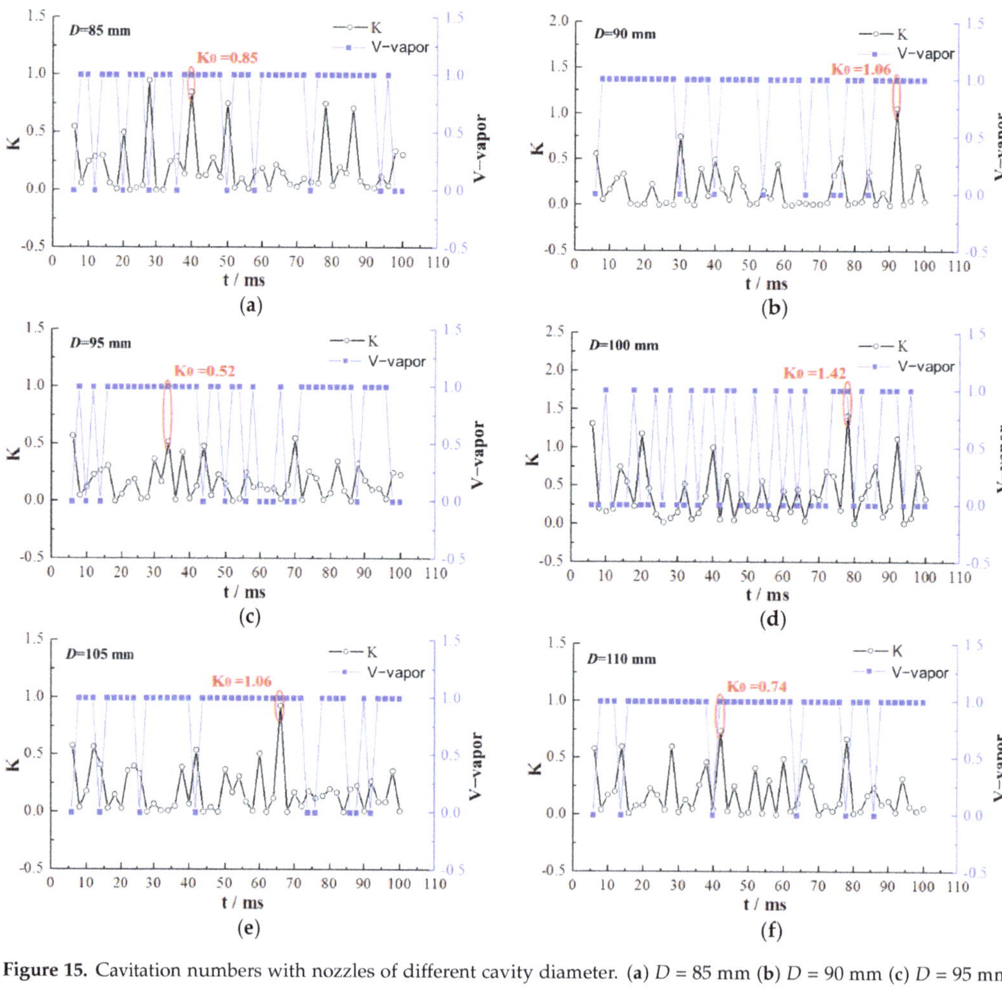

Figure 15. Cavitation numbers with nozzles of different cavity diameter. (**a**) D = 85 mm (**b**) D = 90 mm (**c**) D = 95 mm (**d**) D = 100 mm (**e**) D = 105 mm and (**f**) D = 110 mm.

3.1.3. Influence of Cavity Length on Jet Performance

1. Influence of cavity length on peak velocity and oscillation frequency

The jet performance of the nozzle was analysed using cases with the structural parameters of 9-100-60-L-11 (where L = 25 mm, 30 mm, 35 mm, 40 mm, 45 mm and 50 mm respectively, and d_1 = 9 mm, D = 100 mm, α = 60°, d_2 = 11 mm). Figure 16 shows that as the cavity length L increases, the peak velocity and feedback velocity first increase and then decrease. At L/d_2 values of 3–3.5, the peak velocity and feedback velocity are both higher. As L/d_2 increases from 2.27 to 3.18, the disturbance wave develops more and more fully in the disturbance layer, and the feedback velocity of the disturbance wave increases gradually. However, as the further increase of L/d_2, when L/d_2 > 3.18, the path of the disturbance wave increases, and more energy is spread to the vortex ring. It leads to the gradual decrease of the feedback velocity.

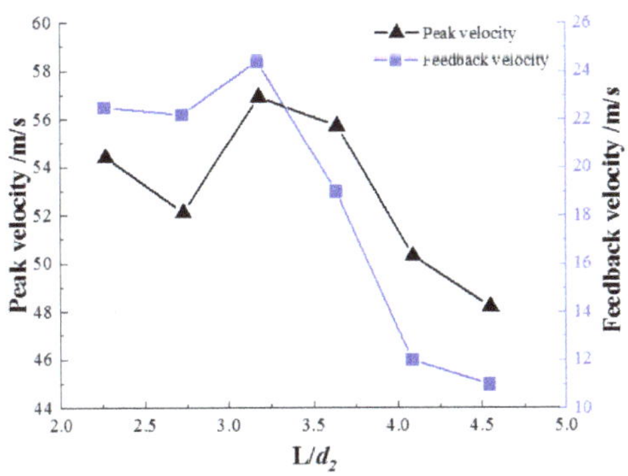

Figure 16. Curves of peak velocity and feedback velocity with cavity length.

Figure 17 shows that the steam pocket region is affected by the length of the cavity. When the cavity length increases, the steam pocket gradually increases. However, if the cavity length is too short, the disturbance wave cannot fully develop in the shear layer, which prevents the feedback velocity from reaching its peak.

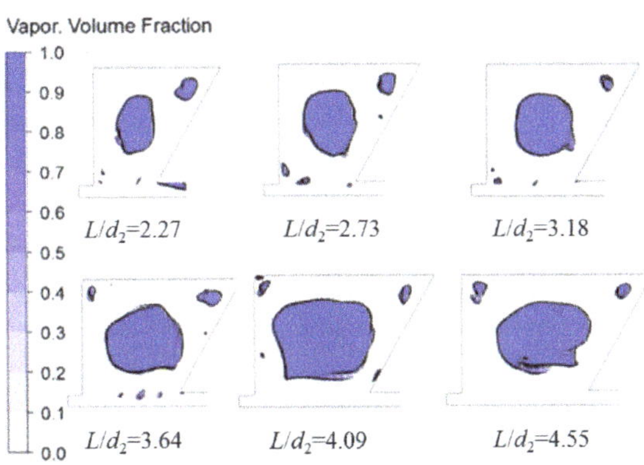

Figure 17. Steam pocket regions of the nozzle at different cavity lengths.

2. Influence of cavity length on cavitation number

Figure 18 shows the curves of the vapour volume fraction and cavitation number with time. The figure demonstrates that the critical cavitation number peaks at a cavity length of 35 mm.

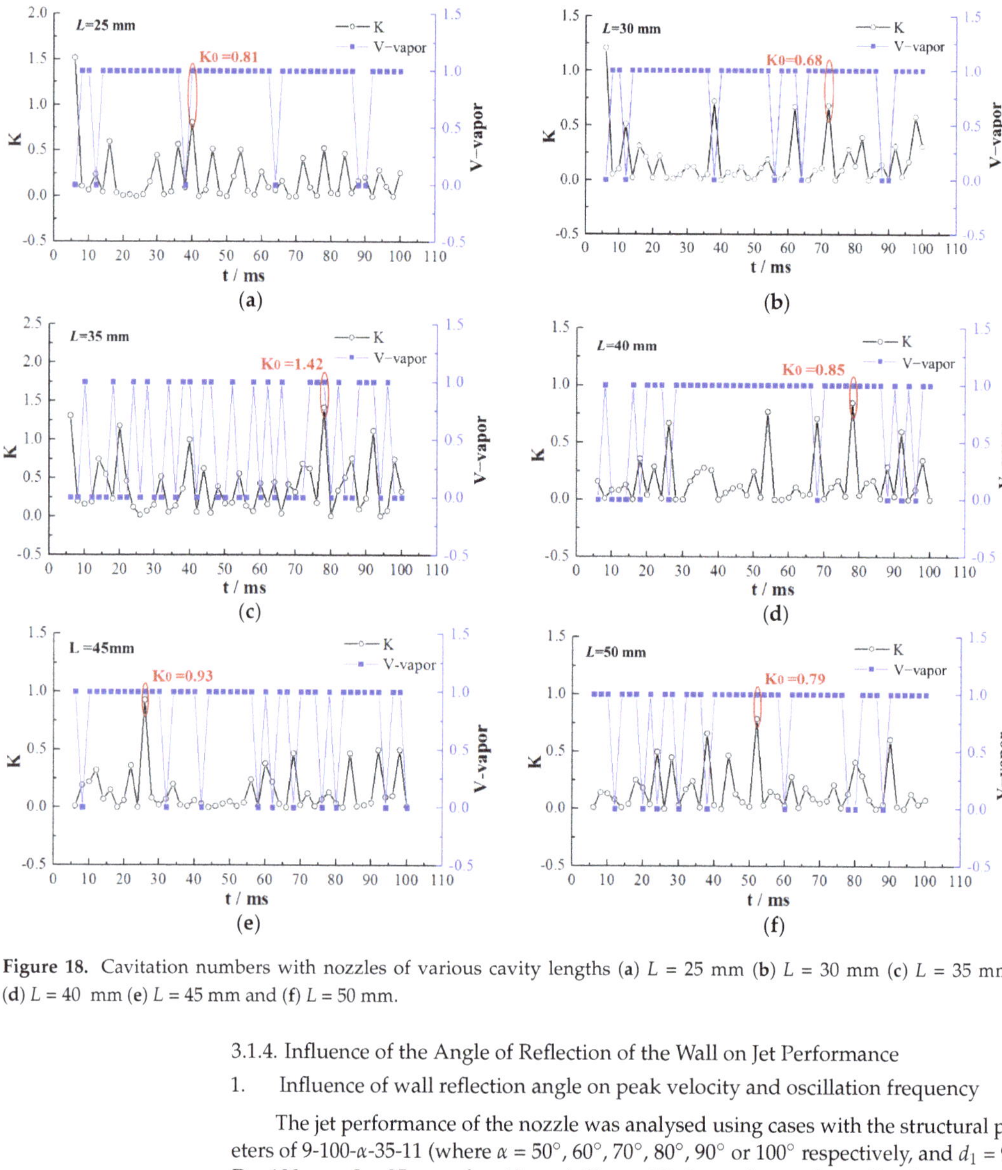

Figure 18. Cavitation numbers with nozzles of various cavity lengths (**a**) L = 25 mm (**b**) L = 30 mm (**c**) L = 35 mm (**d**) L = 40 mm (**e**) L = 45 mm and (**f**) L = 50 mm.

3.1.4. Influence of the Angle of Reflection of the Wall on Jet Performance

1. Influence of wall reflection angle on peak velocity and oscillation frequency

The jet performance of the nozzle was analysed using cases with the structural parameters of 9-100-α-35-11 (where α = 50°, 60°, 70°, 80°, 90° or 100° respectively, and d_1 = 9 mm, D = 100 mm, L = 35 mm, d_2 = 11 mm). Figure 19 shows that as the wall reflection angle increases, the peak velocity first increases and then decreases. The peak velocity is maximum when α = 60°. The feedback velocity increases with the increase in α, and becomes stable when α is greater than 60°. This is because when α increases from 50° to 60°, the low-speed fluid in the oscillating cavity gradually decreases. Under the same inlet jet energy, the low-speed fluid consumes less disturbed wave energy, and the feedback velocity of the disturbed wave gradually increases. When α is greater than 60°, although the low-velocity fluid in the oscillating cavity decreases gradually, the steam pocket in the oscillating cavity becomes larger significantly as shown in Figure 20. The larger steam pocket will absorb more energy of the disturbed wave. Under the comprehensive function of the two factors, the feedback velocity of the disturbed wave gradually becomes stable.

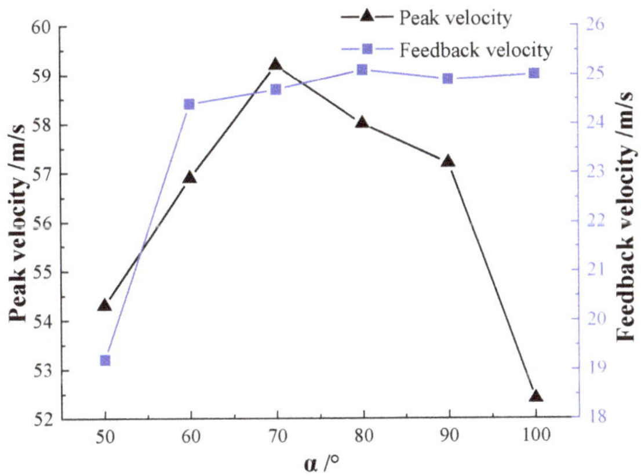

Figure 19. Curves of peak velocity and feedback velocity with wall reflection angle.

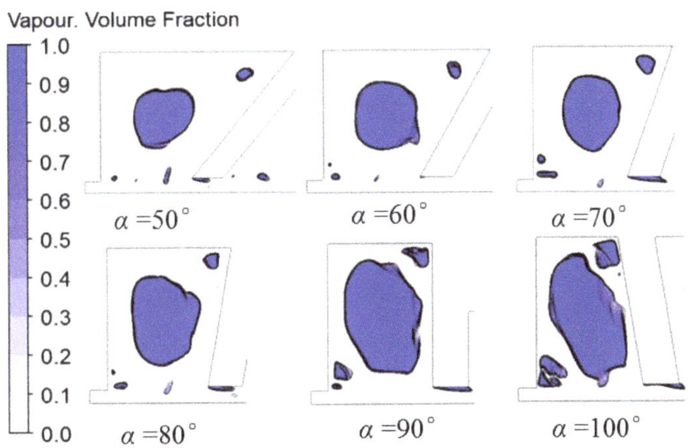

Figure 20. Steam pocket regions of the nozzle with different wall reflection angles.

Figure 20 shows that the steam pocket region is affected by the wall reflection angle. As the angle increases, the steam pocket increases gradually and the serial bubbles decrease. Therefore, the oscillation frequency gradually decreases with increases in angle.

2. Influence of wall reflection angle on cavitation number

Curves of vapour volume fraction and cavitation number with time are shown in Figure 21. The critical cavitation number peaks at an angle of 60°.

3.2. *Synthesis of Results and the Influence of Operating Parameters on Jet Performance*

1. Synthesis of results

By analysing the influences of the structural parameters on jet performance, it was found that the peak velocity is greatest when the value of d_1/d_2 is 0.818. The cavitation capability is the strongest when d_1/d_2 = 0.818. Considering the cleaning requirements, the optimal value of d_1/d_2 is 0.818. When the value of D/d_2 is 9.09, the nozzle has better cavitation capability, a better oscillation effect, and a greater peak velocity. The peak velocity and feedback velocity are highest when L/d_2 = 3–3.5. The critical cavitation number peaks when L/d_2 = 3–3.5. The nozzle has a good cavitation capability and oscillation effect at a wall reflection angle of 60°. At 70°, the peak velocity reaches its maximum value but the cavitation capability is very weak. Therefore, the angle should be set to 60–70°.

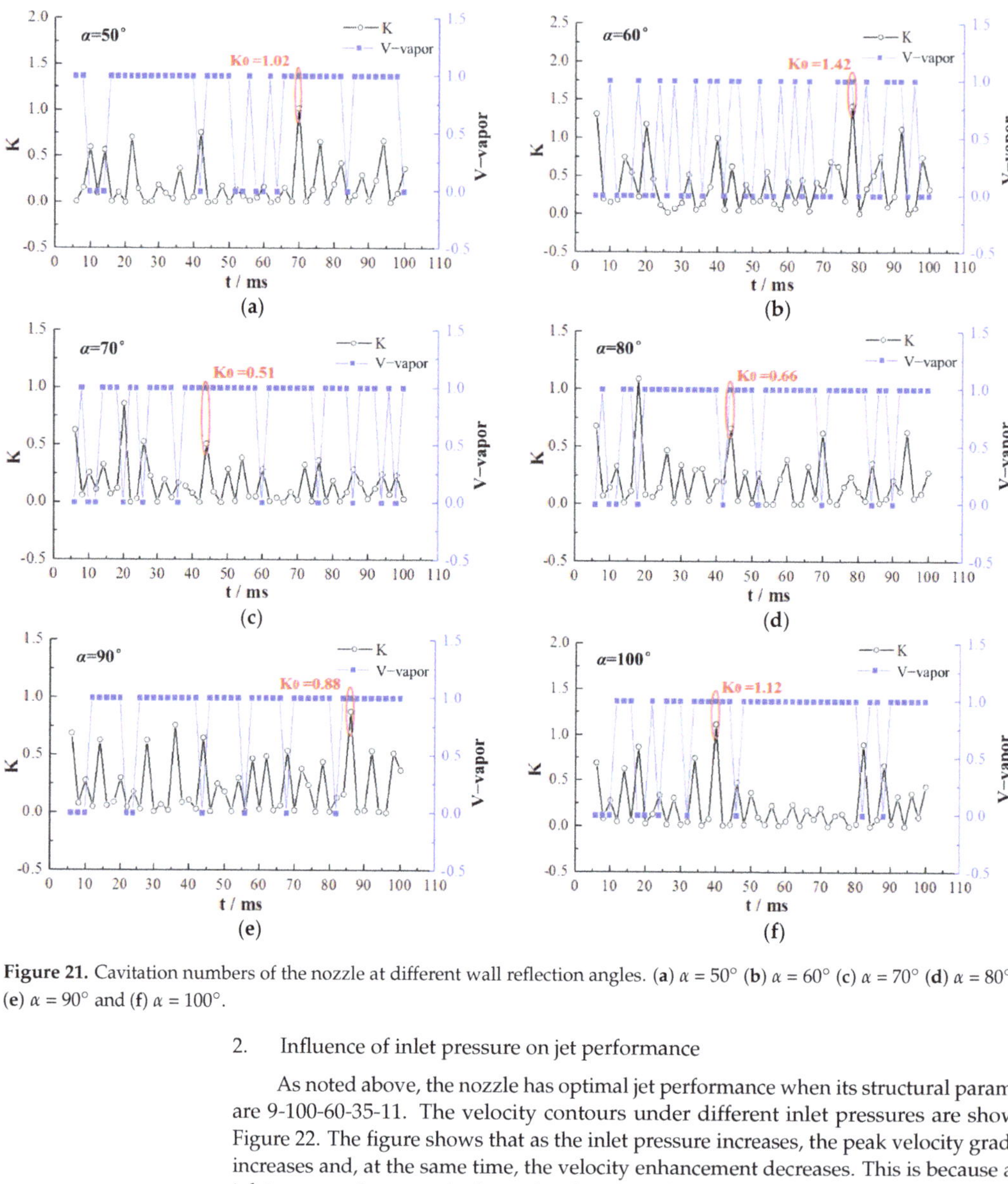

Figure 21. Cavitation numbers of the nozzle at different wall reflection angles. (**a**) $\alpha = 50°$ (**b**) $\alpha = 60°$ (**c**) $\alpha = 70°$ (**d**) $\alpha = 80°$ (**e**) $\alpha = 90°$ and (**f**) $\alpha = 100°$.

2. Influence of inlet pressure on jet performance

As noted above, the nozzle has optimal jet performance when its structural parameters are 9-100-60-35-11. The velocity contours under different inlet pressures are shown in Figure 22. The figure shows that as the inlet pressure increases, the peak velocity gradually increases and, at the same time, the velocity enhancement decreases. This is because as the inlet pressure increases in the cavity, there is an increase in the loss of kinetic energy, which leads to a gradual reduction in the peak velocity enhancement. After this, the disturbance wave and its velocity are rapidly increased and there is an increase in the inlet pressure. As the inlet pressure increases, the number of the disturbance wave increases significantly and its velocity increases. Then the disturbance frequency of the disturbance wave on the axis velocity increases, which finally results to the increase of the oscillation frequency of the pulsed jet.

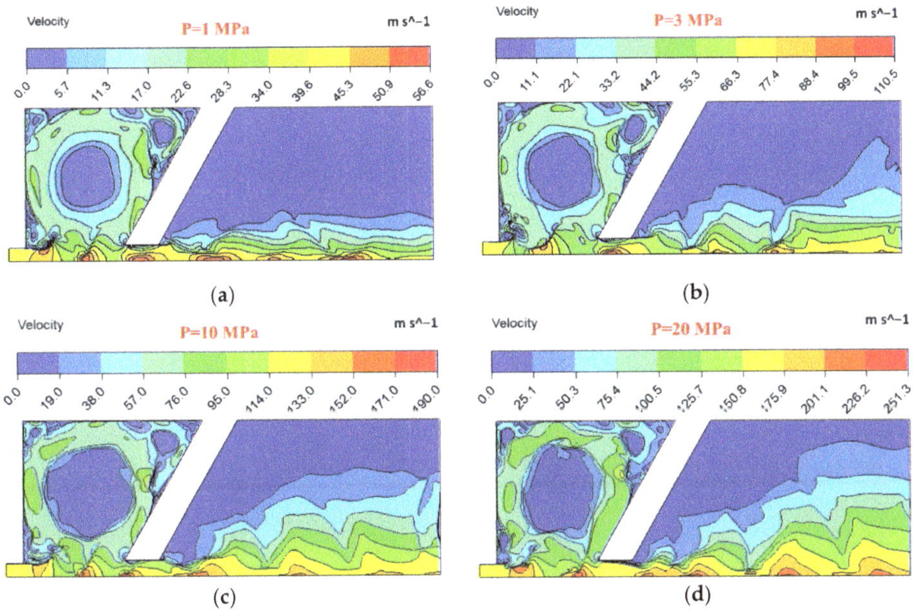

Figure 22. Velocity contours of the nozzle at different inlet pressures. (**a**) P = 1 MPa (**b**) P = 3 MPa (**c**) P = 10 MPa and (**d**) P = 20 MPa.

4. Conclusions

This paper first analysed the factors influencing jet performance. Then, the flow field of the nozzle was analysed by numerical simulation, the results of which had high consistency with the experimental results. Finally, to determine the nozzle with optimal oscillation and cavitation capability, the influences of structural and operating parameters on jet performance were analysed. The following conclusions were obtained via analysis of the models:

(1) A self-excited oscillating pulse jet formation mechanism was described. The results show that the number of serial eddies in the disturbance layer is proportional to the oscillation frequency. When the serial eddies remain the same, there is a larger steam pocket and lower oscillation frequency.

(2) The influences of the key structural parameters of self-excited oscillating nozzles on peak velocity, oscillation frequency, and critical cavitation number were revealed. With increases in inlet diameter, cavity diameter, cavity length, and wall reflection angle, the oscillation frequency decreases. Peak velocity fluctuates with changes in the cavity diameter, while the critical cavitation number increases with cavity diameter. As the inlet pressure increases, the peak velocity gradually increases and the peak velocity enhancement decreases.

(3) The working performance of a self-excited oscillating nozzle is influenced by its structural parameters and working pressure. The nozzle has a better cavitation capability, oscillation effect and greater peak velocity when d_1/d_2 = 0.818, D/d_2 is approximately 9, L/d_2 = 3–3.5, and the angle α= 60–70°. Then, in this situation, the self-excited oscillating nozzle has better cleaning ability at a constant working pressure. However, the increase in working pressure is beneficial to increasing the peak velocity and enhancing the cleaning ability of the self-excited oscillating nozzle.

Author Contributions: Conceptualization, methodology, and validation, B.F.; formal analysis and writing—original draft preparation, S.Z.; writing—review and editing, L.S. All authors have read and agreed to the published version of the manuscript.

Funding: This project was supported by the Natural Science Foundation of China (No. 51974036 and No. 51604039), the Yangtze Fund for Youth Teams of Science and Technology Innovation (No. 2016cqt01) and the New Generation Information Technology Innovation Program (No. 2019ITA04001).

Institutional Review Board Statement: Not applicable.

Informed Consent Statement: Not applicable.

Data Availability Statement: The data that support the finding of this study are available within the article.

Conflicts of Interest: The authors declare that they have no conflict of interest.

Nomenclature

ν	Viscosity coefficient, dimensionless, $\nu = 1/\text{Re}$
Re	Reynolds number, dimensionless
F	Body force, N
P	Fluid pressure, Pa
u	Velocity, m/s
Δ_i	Length of grid, m
τ_{ij}	Grid dimension stress, N
v_T	Eddy viscosity coefficient, dimensionless
Cs	Smagorinsky constant, $Cs = 0.1$
S_{ij}	Tensor of deformation rate
Δ	Filtering width, m, $\Delta = (\Delta x_2 + \Delta y_2)^{1/2}$
ρ	Mixture density, kg/m^3
μ	Mixture dynamic viscosity, Pa·s
$\vec{v}$	Velocity, m/s
t	Time, s
n	Number of bubbles
r	Radius of the bubble, m
σ	Surface tension of the fluid, N
p_{vap}	Pressure within the bubble, Pa
D	Cavity diameter, m
L	Cavity length, m
α	Reflection angle of wall, degree
d_0	Pipe diameter, m
a	Wave velocity, m/s
K	Cavitation number, dimensionless
p_i	Absolute pressure of fluid, Pa
p_s	Saturation pressure of fluid, Pa
Δx, Δy	The width of grid along with the X axis and Y axis direction, m
ρ_l, ρ_v	Liquid density, Vapour density, kg/m^3
μ_L, μ_v	Liquid dynamic viscosity, Vapour dynamic viscosity, Pa·s
β_l, β_v	Volume fraction of liquid, Volume fraction of vapour, %
d_1, d_2	Inlet diameter, Outlet diameter, m

References

1. Domínguez, A.; Menéndez, J.A.; Inguanzo, M.; Pis, J.J. Investigations into the characteristics of oils produced from microwave pyrolysis of sewage sludge. *Fuel Process. Technol.* **2005**, *86*, 1007–1020. [CrossRef]
2. Han, W.L. Storage Tank Corrosion and Its Protection Measures. *Pet. Eng. Constr.* **2010**, *36*, 41–47.
3. Xu, R.L. Present Situation and Research of Treatment of Sludge from Oil-tanks. *Petrochem. Saf. Technol.* **2003**, *19*, 36–39.
4. Shaheen, S.E.; Zhou, A.L. Chemical processing and machinery operation of tank cleaning. *Foreign Oilfield Eng.* **2000**, *7*, 48–50.
5. Wartel, M. Method for Cleaning an Oil Storage Tank and Device for Implementing Same. European Patent EP1091812, 22 August 2007.

6. Dixon, R. Storage Tank Cleaning Method and Apparatus. U.S. Patent EP2303475, 23 December 2009.
7. Morel, T. Experimental study of a jet-driven Helmholtz oscillator. *Am. Soc. Mech. Eng.* **1978**, *101*, 383–390. [CrossRef]
8. Liao, Z.F.; Tang, C.L. Theory of the self-excited oscillation pulsed jet nozzle. *J. Chongqing Univ.* **2002**, *25*, 24–27.
9. Liao, Z.F.; Li, J.; Chen, D.; Deng, X.; Tang, C.; Zhang, F. Theory and experimental study of the self-excited oscillation pulsed jet nozzle. *Chin. J. Mech. Eng.* **2003**, *16*, 379–383. [CrossRef]
10. Fang, Z.L.; Zeng, F.D.; Xiong, T.; Wei, W.; Jiang, P.; Wu, Q.; Wang, Y.; Fei, Y. Large eddy simulation of self-excited oscillation inside Helmholtz oscillator. *Int. J. Multiph. Flow* **2020**, *126*, 103253. [CrossRef]
11. Liu, X.Y.; Xu, H.Z.; Zhao, L.; Yu, X.; Chen, H.; Zhang, S.; Ji, J. Investigation of the impact characteristics and pulse mechanism of a self-excited aspiration pulsed jet device. *Exp. Therm. Fluid Sci.* **2021**, *124*, 110371. [CrossRef]
12. Wang, H.M.; Jiao, L. Numeireal simulation of self- excited oscillation Pulsed jet and analysis of Parameters' influence. *J. Zhejiang Univ. (Eng. Sci.)* **2005**, *39*, 1450–1454.
13. Hu, D.; Li, X.H.; Tang, C.-L.; Kang, Y. Analytical and experimental investigations of the pulsed air–water jet. *J. Fluids Struct.* **2015**, *54*, 88–102. [CrossRef]
14. Tang, C.L.; Hu, D.; Zhang, F.H. Study on the frequency characteristic of self-excited oscillation pulsed water jet. *Adv. Mater. Res.* **2011**, *317–319*, 1456–1461. [CrossRef]
15. Peng, G.Y.; Shimizu, S. Progress in numerical simulation of cavitation water jets. *J. Hydrodyn.* **2013**, *25*, 502–509. [CrossRef]
16. Tamaki, N.; Shimizu, M.; Hiroyasu, H. Enhancement of the atomization of a liquid jet by cavitation in a nozzle hole. *At. Sprays* **2001**, *11*, 125–137.
17. Payri, F.; Payri, R.; Salvador, F.J.; Martínez-López, J. A contribution to the understanding of cavitation effects in Diesel injector nozzles through a combined experimental and computational investigation. *Comput. Fluids* **2012**, *58*, 88–101. [CrossRef]
18. Xiang, L.H.; Wei, X.S.; Chen, S.Y. Experimental study on the frequency characteristics of self-excited pulsed cavitation jet. *Eur. J. Mech. B/Fluids* **2020**, *83*, 66–72. [CrossRef]
19. Huang, M.; Kang, Y.; Wang, X.C.; Hu, Y.; Cai, C.; Liu, Y.; Chen, H. Experimental investigation on the rock erosion characteristics of a self-excited oscillation pulsed supercritical CO_2 jet. *Appl. Therm. Eng.* **2018**, *139*, 445–455. [CrossRef]
20. Li, H.S.; Liu, S.Y.; Jia, J.G.; Wang, F.; Guo, C. Numerical simulation of rock-breaking under the impact load of self-excited oscillating pulsed waterjet. *Tunn. Undergr. Space Technol.* **2020**, *96*, 103179. [CrossRef]
21. Egerer, C.P.; Hickel, S.; Schmidt, S.J.; Adams, N. Large-eddy simulation of turbulent cavitating flow in a micro channel. *Phys. Fluids* **2014**, *26*, 1–40. [CrossRef]
22. Ji, B.; Luo, X.W.; Arndt, R.E.; Peng, X.; Wu, Y. Large Eddy Simulation and theoretical investigations of the transient cavitating vortical flow structure around a NACA66 hydrofoil. *Int. J. Multiph. Flow* **2015**, *68*, 121–134. [CrossRef]
23. Ji, B.; Luo, X.W. Three-dimensional large eddy simulation and vorticity analysis of unsteady cavitating flow around a twisted hydrofoil. *J. Hydrodyn.* **2013**, *25*, 510–519. [CrossRef]
24. Molina, S.; Salvador, F.J.; Carreres, M.; Jaramillo, D. A computational investigation on the influence of the use of elliptical orifices on the inner nozzle flow and cavitation development in diesel injector nozzles. *Energy Convers. Manag.* **2014**, *79*, 114–127. [CrossRef]
25. Vinkovic, I.; Aguirre, C.; Simoë, S.; Gorokhovski, M. Large eddy simulation of droplet dispersion for inhomogeneous turbulent wall flow. *Int. J. Multiph. Flow* **2006**, *32*, 344–364. [CrossRef]
26. Yu, Y.S.; Li, G.X. Multiphase Large Eddy Simulation of Diesel Fuel Spray Atomization. *Chin. Intern. Com-Bustion Engine Eng.* **2009**, *30*, 39–44.
27. Huang, B.; Zhao, Y.; Wang, G.Y. Large Eddy Simulation of turbulent vortex-cavitation interactions in transient sheet/cloud cavitating flows. *Comput. Fluids* **2014**, *92*, 113–124. [CrossRef]
28. Li, G.S.; Shen, Z.H. *Theory and Application of Self-Resonating Cavitating Water Jet*; Chain University of Petroleum Press: Dongying, China, 2008.
29. Lai, S.Q.; Liao, Z.F. The Theory and Experimental Study of the Self-Excited Oscillation Pulsed Jet Nozzle (Pipeline Pulsed Flow Generator). *Nat. Resour.* **2013**, *4*, 395–403. [CrossRef]
30. Zhang, Z.S.; Cui, G.X.; Xu, C.X. *Theory and Application of Numerical Simulation of Large Eddy Simulation Turbulence*; Qinghua University Press: Beijing, China, 2008.
31. Du, Y.K.; Wang, R.H.; Ni, H.J. Large eddy simulation of self-oscillation pulsed water jet drawing in annulus fluid. *J. Hydrodyn. (Ser. A)* **2009**, *24*, 455–462.
32. Li, H.S. Rock Breaking Performance of Self-Excited Oscillating Pulsed Waterjet. Ph.D. Thesis, China University of Mining and Technology, Beijing, China, 2020.
33. Wang, P.; Ni, H.J.; Wang, R.H. Modulating downhole cuttings via a pulsed jet for efficient drilling-tool development and field testing. *J. Nat. Gas Sci. Eng.* **2015**, *27*, 1287–1295. [CrossRef]
34. Keller, A.P. Cavitation Scale Effects—Empirically Found Relations and the Correlation of Cavitation Number and Hydrodynamic Coefficients. In Proceedings of the CAV 2001: Fourth International Symposium on Cavitation, California Institute of Technology, Pasadena, CA, USA, 20–23 June 2001.

processes

MDPI

Article

Risk Classification of Shale Gas Gathering and Transportation Pipelines Running through High Consequence Areas

Kun Chen [1,*], Nan Shi [1], Zhenjie Lei [1], Xu Chen [1], Wei Qin [2], Xin Wei [1] and Shanghao Liu [3,*]

1 School of Safety Engineering, Chongqing University of Science and Technology, Chongqing 401331, China; shinan207014@163.com (N.S.); lei_zhenjie@163.com (Z.L.); chenxv1895@163.com (X.C.); wxin2018@126.com (X.W.)
2 Chongqing Gas District, PetroChina Southwest Oil and Gasfield Company, Chongqing 400021, China; qwxr050106@163.com
3 School of Chemical Engineering, Anhui University of Science and Technology (AUST), Huainan 232001, China
* Correspondence: ckym117@cqust.edu.cn (K.C.); shliu998@163.com (S.L.)

Abstract: Shale gas gathering and transportation pipeline poses significant risk due to special geographical conditions and different climatic conditions in high consequence areas such as Sichuan and Chongqing. The risks become critical as gas pipelines run through high consequence areas such as hospital, market, and scenic areas. This study presents a risk classification method for the pipelines running through high consequence areas. The proposed method considers different failure scenarios including third-party damage, corrosion, design and construction defects, mis-operation, and natural disasters. The method uses subjective and objective data from different sources. To minimize the subjectivity and data uncertainty, an improved fuzzy analytic hierarchy process was used to process data. The estimated risk is used to classify different risk zones. After the failure of shale gas pipelines in HCAs, in order to reduce the adverse impact of emergencies, personnel should immediately organize an evacuation to a safe area, focusing on the diagnosis and analysis of risk factors that are more likely to lead to pipeline leakage. The developed classes are verified using field data. The study observes that risk levels classified using the proposed method provide realistic assessments of hazard zoning. Risk zoning will help develop effective risk management strategies.

Keywords: pipeline safety; risk classification; pipeline failure; fuzzy analytic hierarchy process; pipeline risk assessment; shale gas pipelines

Citation: Chen, K.; Shi, N.; Lei, Z.; Chen, X.; Qin, W.; Wei, X.; Liu, S. Risk Classification of Shale Gas Gathering and Transportation Pipelines Running through High Consequence Areas. *Processes* **2022**, *10*, 923. https://doi.org/10.3390/pr10050923

Academic Editor: Jean-Claude Assaf

Received: 12 April 2022
Accepted: 4 May 2022
Published: 6 May 2022

1. Introduction

China has a high accident rate in oil and gas pipelines compared to European and American countries. According to statistics, there have been over 1000 pipeline safety accidents since 1995 [1]. The pipeline accidents in High Consequence Areas (HCAs) will have severe impact. The pipeline is considered as main mode of transportation. As per 2017 data, oil and gas pipelines reached 133,100 km in China [2]. Adding each kilometer of pipeline increases potential risks, thus a detailed methodology to analyze and characterize the risk of potential pipeline accidents is needed more importantly in high consequence areas.

HCAs are classified based on geographical conditions, population, quality of pipeline ontology, and other factors. The United States was one of the first countries to study HCAs in terms of development [3]. In 1988, Onisawa first proposed to transform Fuzzy Possibility Score (FPS) into Fuzzy Possibility (FP) in order to calculate pipeline risks in HCAs [4]. In 2016, Lam and Zhou studied the distribution of pipeline accidents in the United States based on database, including installation year, regional grade, failure cause, and other parameters, and their research laid a foundation for the quantitative risk assessment of pipeline HCAs [5]. Compared with foreign countries, domestic research on HCAs started late, but it also has made some achievements. In 2010, Zhang Peng et al. proposed a multi-hierarchy grey relational analysis method for oil and gas pipelines for the first time; however, they did not specifically point out HCAs [6]. In 2015, Wang Xiaolin et al. proposed to divide HCAs into three types for the first time [7], namely population density, important facilities, and environmental sensitivity. In recent years, a quantitative evaluation model and scoring index for classification of HCAs were developed. These method attributes include distance, vulnerability of the region, ecological sensitivity, population density, and the number of other sensitive receptors. In 2014, Dong Shaohua et al. conducted a systematic study on upgrading regional grades in China. They started from pipeline risks and learned from the laws and regulations on upgrade management for foreign pipeline companies and proposed measures to be taken after upgrading. However, there was no specific study on upgrade management and risk assessment methods of pipeline regional grades [8]. In 2016, Shan Ke and Shuai Jian applied three-stage process evaluation methods to the upgraded gas transmission pipeline in the region and put forward the risk management measures for regional upgraded gas transmission pipelines from two aspects, including technical transformation and regular maintenance [9]. In 2017, Yao Anlin et al. formulated corresponding management procedures for gas pipelines with different risk levels in combination with the regional grade change conditions of gas transmission pipelines. They established a risk assessment model for natural gas pipelines in upgraded areas. According to the possibility of upgrading gas pipelines, relevant principles for risk control were proposed [10]. Based on relevant domestic research, the study of HCAs has lagged behind that of foreign countries for at least ten years. The formulation of relevant standard and specifications mostly refers to the foreign standard system, which is inconsistent with the domestic current situation, the adoption rate of standards is low, and there are deviations in the actual application process. The above problems have become the key factors restricting pipeline integrity management in China. Therefore, for research on HCAs of shale gas gathering and transmission pipeline, hierarchical management should be carried out under the guidance of national and enterprise standards and specifications combined with regional current situation and geographical characteristics, which are very important for a unified management of HCAs and pipeline integrity management.

China is focusing on the identification and classification of HCAs areas. As a new generation of gas reservoirs is used, there are limited studies on shale gas pipelines running through HCAs. The shale gas gathering and transportation pipelines are prone to accidents due to their complex operation and severe operating conditions. This study presents a risk classification method for the pipelines running through high consequence areas. The study will help risk management and pipeline integrity management in HCAs.

2. Materials and Methodology

An improved fuzzy analytic hierarchy process (IFAHP) introduces the weight coefficient and triangular fuzzy number of experts and extends Analytic Hierarchy Processes (AHP) to the field of group decision-making and fuzzy decision making. The calculation process of IFAHP includes three parts. It establishes the evaluation index set and then determines each index weight to obtain the comprehensive weight. Finally, a fuzzy comprehensive evaluation of pipeline risk grades in HCAs is performed. The specific process of HCAs classification is shown in Figure 1.

(1) HCAs identification: determine the pipelines running through HCAs to be analyzed;
(2) Establish pipelines failure index system: identify different failure scenarios of shale gas gathering and transmission pipelines and record the hierarchical relationship between each failure factor;
(3) Single factor analysis: according to the secondary index of shale gas pipeline failure factors in HCAs and the principle of failure probability scoring, determine the membership degree of each single factor;
(4) Determine index objective weight: combine with expert scoring and transform the failure model of shale gas pipelines into Bayesian network. Minimize subjectivity and data uncertainty;

(5) Fuzzy comprehensive evaluation of HCAs: according to the weight of each index obtained in the above steps to determine the comprehensive weight. It can be used for risk classification and the evaluation of HCAs.

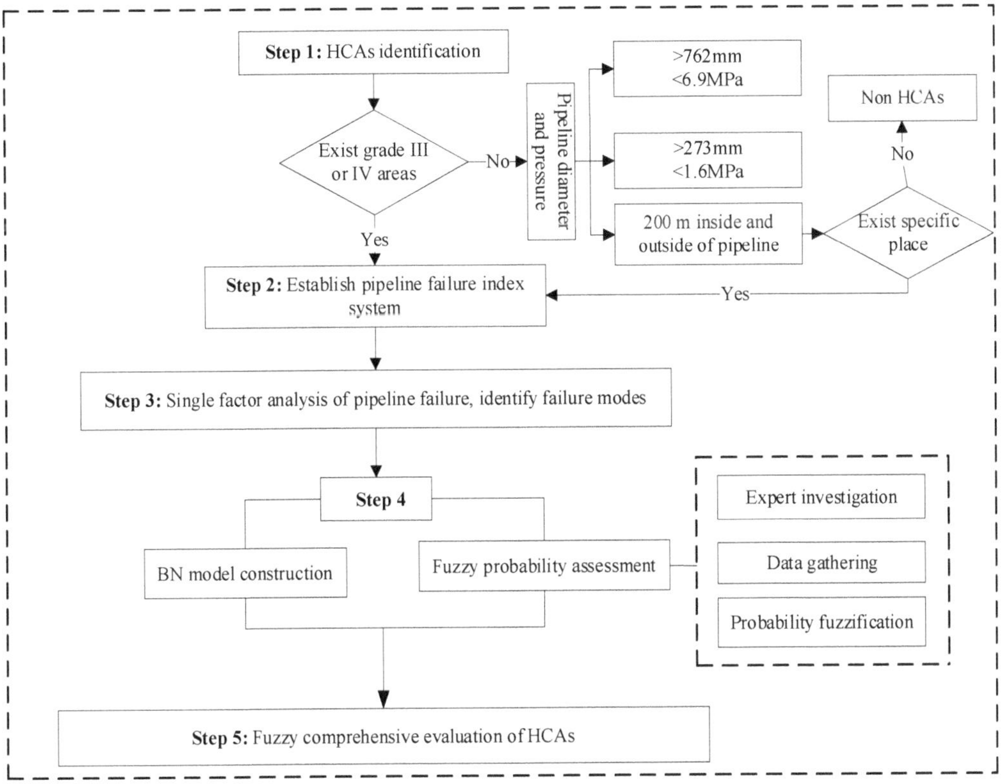

Figure 1. Hierarchical management flow chart of HCAs.

2.1. Establish Evaluation Index Set

The total objective A is divided into m subindex sets, and their relationship meets the following conditions:

$$A = B_1, B_2 \cdots B_m, B_i \cap B_j \neq \varnothing, i \neq j \tag{1}$$

where B is the subindex set, and they are independent of each other; m, i, and j are the number of subindexes.

Use the nine-scale scoring principle combined with the triangular fuzzy number method to conduct fuzzy judgement of the index at all levels; the results are shown in Table 1.

Table 1. Indicator scale table.

Scale	Definition
1	Very unimportant
3	Slightly unimportant
5	General important
7	Slightly important
9	Very important
3, 4, 6, 8	Between 1–2 or 3–5 or 5–7 or 7–9

Assuming that B is the parent index and that it has a membership relationship with $C_1, C_2, \cdots, C_n$, expert scores according to the index scale, the fuzzy judgement matrix R_w is obtained, which is express as follows.

$$R_w = [r_1 r_2 \ldots r_n]_{1\times n} \tag{2}$$

In this expression, $r_i = (l_i, m_i, u_i)$, which is triangular fuzzy number, $i = 1, 2, \cdots, n$. In comparison between C_i and superior B, li is the most pessimistic estimate, m_i is the most probable estimate, and u_i is the most optimistic estimate. R_w represents the importance of the sub-indicator, which is a $1 \times n$ triangular fuzzy matrix.

2.2. Determine Each Index Weight

IFAHP is used to determine the importance weight of each index. To calculate the subjective weight of factors, the event statistical objective weight is combined to obtain the comprehensive weight of each factor. This method avoids the problem of traditional AHP that is dependent on the experience of the expert and the difficulty of adjusting the consistency of the judgment matrix. This method reflects the fuzzy decision making of AHP. This method introduces the expert weight coefficient, combines fuzzy triangle number with AHP, and modifies it with a Bayesian network. By minimizing subjectivity and data uncertainty, IFAHP is used to process data and also improves consistency by transforming the possibility of each index into real values to solve weight. This method not only improves practicability, but also solves the problem of judgement matrix inconsistency in AHP.

2.2.1. Determine Expert Scoring Weight

Traditional AHP relies too much on the experience level of experts, and it is difficult to adjust the consistency of the judgment matrix; thus, IAHP is used to calculate the expert rating weight of each factor [11], which is marked as α_i. The method extends AHP to the field of expert group strategy and takes full factors into account, such as expert's personal ability, experience, and level; thus, this minimizes subjectivity. In addition, by converting the probability degree matrix of each index real value, the weight was solved, and the steps of adjusting the consistency of judgment matrix by traditional AHP were optimized. According to the scoring results of several different experts, combined with different expert information, the comprehensive scoring value of each index is calculated and normalized, as is shown in Appendix A. The expression of the expert weight coefficient is as follows.

$$G_k = a_k \times b_k \times c_k \times d_k \times e_k \tag{3}$$

$$\beta_k = G_k / \sum_{k=1}^{s} G_k \tag{4}$$

In the expression, a_k is expert popularity, b_k is professional title, c_k is educational background, d_k is problem familiarity, e_k is evaluation confidence, and the expert weight coefficient score is shown in Table 2.

Table 2. Expert weight coefficient score table.

Number	Indicator	Expert Category	Score
1	Expert popularity	National famous scholar, provincial and ministerial scholar, other	3, 2, 1
2	Professional title	Associate senior and above, intermediate, other	3, 2, 1
3	Educational background	Doctor, master's degree, undergraduate	3, 2, 1
4	Problem familiarity	Major, specialty-related, not related to the major	3, 2, 1
5	Evaluation confidence	Confident, less confident, general confident	3, 2, 1

In order to compare the proximity of two triangular fuzzy numbers, the possibility degree of a triangular fuzzy number is defined. The probability degree is converted into a real numerical judgement matrix as follows:

$$V(M_2 \geq M_1) = \mu_{M_2}(d) = \begin{cases} 1 & b_2 \geq b_1 \\ 0 & a_1 \geq c_2 \\ (a_1 - c_2)/[(b_2 - c_2) - (b_1 - a_1)] & other \end{cases} \tag{5}$$

where $M_1 = (a_1, b_1, c_1)$ and $M_2 = (a_2, b_2, c_2)$; they are two arbitrary triangular fuzzy numbers.

According to formula (5), the possibility value of $C_1, C_2, \cdots C_n$ is calculated. Then, make pairwise comparisons to obtain V, and the expression is as follows:

$$V = (\gamma_{ij})_{n\times n} = \begin{bmatrix} 1 & V(\tilde{S_1} \geq \tilde{S}_2) & \cdots & V(\tilde{S_1} \geq \tilde{S}_n) \\ V(\tilde{S_2} \geq \tilde{S}_1) & 1 & \cdots & V(\tilde{S_2} \geq \tilde{S}_n) \\ \vdots & \vdots & \ddots & \vdots \\ V(\widetilde{S_n} \geq \tilde{S}_1) & V(\widetilde{S_n} \geq \tilde{S}_2) & \cdots & 1 \end{bmatrix} \tag{6}$$

where V is the possible degree matrix, and $\widetilde{S_i}$ is normalized comprehensive judgement matrix.

2.2.2. Determine Each Index Weight

Bayesian network is a graphical model of probability, based on Bayesian formula. By constructing fault tree and then transforming it into a Bayesian network, it makes up for the limitation of quantitative analysis of fault trees [12]. A Bayesian network can use the Bayesian theory and new information about event occurrence to update the failure probability of events, realizing the dynamic analysis of the entire system. BN can be expressed by using B = <N, P>, where N represents a structure graph constructed by the network nodes having causality, and P represents the probability distribution of the nodes [13].

Combined with the pipeline failure index model, the mapping and logic relationship are unchanged by constructing the evaluation factor model and then transforming it into a Bayesian network. On the basis of the prior probability of basic events and the failure probability of shale gas pipelines, probabilistic inference on the prior probability of all remaining non-root nodes in the accidental Bayesian network is performed. The probability of shale gas pipelines leakage accidents in HCAs and the probability of each consequence are obtained. It not only makes up for the shortcomings of fault tree methods, but also makes the establishment of a Bayesian network model simple. It has positive significance for shale gas pipeline hierarchical management and control.

2.2.3. Determine Comprehensive Weight

In order for the comprehensive weight to be as close as possible to both sides and not biased to either sides, the comprehensive weight was obtained by optimizing the model based on the principle of minimum identification information [14,15] and finding the overall weight ω_i. The expression is as follows:

$$\omega_i = \frac{\sqrt{\alpha_i \varepsilon_i}}{\sum_{n}^{j=1} \sqrt{\alpha_i \varepsilon_i}} \tag{7}$$

where α_i is expert scoring weight, and ε_i is probability weight. The comprehensive weight vector is $W = [\omega_1, \omega_2, \cdots, \omega_n]^{\mathrm{T}}$.

2.3. Fuzzy Comprehensive Evaluation of Pipeline Risk Grade in High Consequence Areas

2.3.1. Risk Identification of High Consequence Areas

To distinguish the severity of pipeline accidents in HCAs, the grade of HCAs is divided as the classification Chinese Standard [16] and the characteristics of shale gas development areas (mainly in Sichuan and Chongqing). The principle and grades of HCAs used for shale gas gathering and transportation pipelines are shown in Table 3.

Table 3. Identification principle and grade of HCAs.

Category	Pipe Diameter	Precondition	Subitems	Serial Number
HCAs	Greater than 762 mm, and the maximum allowable operating pressure is more than 6.9 MPa (198 m)	Radius of influence area	Particular area 1	II
			Particular area 2	II
			Particular area 3	II
	Less than 273 mm, and the maximum allowable operating pressure is less than 1.6 MPa(35 m)		Particular area 1	I
			Particular area 2	I
			Particular area 3	I
	Other pipe diameters	200 m inside and outside the pipeline	Particular area 1	I
			Particular area 2	I
			Particular area 3	I
	All the pipe diameters		Level 3 areas	II
			Level 4 areas	III

2.3.2. Build Comment Set

When shale gas gathers and is transported in pipelines across HCAs, there are leakage risk factors. The comment set is a set of element evaluation results in factor set U that may leak during the operation of the pipeline, which is usually represented by V. There are two main aspects in the classification of shale gas gathering and transportation pipeline failure risk: First, improve the accuracy of evaluation results; second, the calculation amount should be reasonable. To ensure the objectivity of evaluation, the evaluation principle should be based on reality. Therefore, the commonly used five-grade classification method is adopted in this paper. The risk evaluation set V corresponds to five elements, and the factor grade classification principle is shown in Table 4.

$$V = \{\text{lowest, low, medium, high, highest}\} \tag{8}$$

Table 4. Failure probability scoring principle.

Principle	Risk Description	Grade
The pipeline has experienced similar failures several times a year or is expected to fail within one year.	Highest	V
Similar failure of pipelines occurs every year or failure is expected to occur within 1–3 years.	High	IV
Similar failure of pipelines occurs every year, or failure is expected to occur within 3–5 years.	Medium	III
Similar failure has occurred in the enterprise or is expected to occur within 5–10 years.	Low	II
Similar failures have occurred in the enterprise or are expected to occur after more than 10 years.	Lowest	I

2.3.3. Sub-Factor Evaluation

For a sub-factor that may cause leakage, when shale gas is gathered and the transportation pipeline passes through high consequence areas, according to the principle of the comment set and actual conditions of the pipeline and based on historical data and expert speculation, score each factor. In this manner, the membership degree of the factor to the comment set is determined, and the evaluation membership degree matrix R is established. The sub-factor evaluation fuzzy set R_i can be expressed as follows:

$$R_i = [r_{i1}, r_{i2}, \cdots, r_{i5}],\ i = 1, 2, \ldots, n,\ j = 1, 2, \ldots, 5 \tag{9}$$

where u_i is single factor, and r_{ij} is the grade evaluation degree of u_i.

The sub-factor evaluation fuzzy set of n factors forms R, which can be expressed as follows:

$$R = \begin{bmatrix} R_1 \\ R_2 \\ \vdots \\ R_n \end{bmatrix} = \begin{bmatrix} r_{11} & r_{12} & \cdots & r_{15} \\ r_{21} & r_{22} & \cdots & r_{25} \\ \vdots & \vdots & \ddots & \vdots \\ r_{n1} & r_{n2} & \cdots & r_{n5} \end{bmatrix} \tag{10}$$

where R is a comprehensive membership evaluation matrix.

2.3.4. Multi Factor Fuzzy Comprehensive Evaluation

The sub-factor evaluation result can only explain the influence of specific factor on the evaluation index, and only by comprehensively considering the influence of all factors on the evaluation target can a multi-factor fuzzy comprehensive evaluation be obtained; the specific expression is as follows:

$$B = W^{\mathrm{T}} \times R = [b_1, b_2, \cdots, b_5] \tag{11}$$

where B is the comprehensive evaluation set, W is index weight factor, and R is a single factor evaluation comprehensive membership matrix.

According to the principle of maximum membership degree and obtained comprehensive evaluation set B, determine the comprehensive result v_j corresponding to b_j, which is the largest element in B.

3. Application of the Proposed Methodology

Take the shale gas gathering and transportation pipeline of shale gas in a city as an example. The horizontal length of the pipeline is 31,494.4 m, the real length is 31,891.5 m, the pipe diameter is 508 mm, and the design pressure is 8.5 MPa. By comparison to shale gas gathering pipeline HCAs identification principles and levels (Table 3), the pipeline passes through three HCAs, and the basic situation is shown in Table 5. The proposed above method combined IFAHP and Bayesian network updates to evaluate HCAs and to classify the risk.

Table 5. Basic situation of HCAs.

Serial Number	Feature Description	Length (km)	Grade of HCAs
HCAs1	Primary school and residential area	2.20	III
HCAs2	Township	4.56	III
HCAs3	Crossing large and medium rivers	0.43	I

3.1. *Risk Assessment of Shale Gas Gathering and Transportation Pipeline*

Shale gas gathering and transportation pipeline is faced with many hazards during their operation. According to the European Gas Pipeline Incident Group (EGIG), the main causes of pipeline failures include corrosion, third-party damage, mis-operation, design and construction defects, and natural disasters [17]. If the pipeline risk management measures are not executed thoroughly, it may cause severe accidents. The failure factors of pipeline are shown in Table 6 based on a comprehensive analysis of the failure factors of shale gas gathering and transmission pipelines, referring to the safety evaluation standards at home and abroad, combining each region management experience of shale gas pipelines, and considering pipe body conditions, transmission mediums, the natural environment condition, and artificial factor.

Table 6. The failure factor of pipeline.

No.	Description	No.	Description
X_1	Third party construction	X_{14}	Protective layer performance failure
X_2	Serious pipeline pressure	X_{15}	cathodic protection measures failure
X_3	Artificially vandalism	X_{16}	Stray current interference
X_4	Effusion containing Cl^-, HCO_3^- plasma	X_{17}	Unreasonable pipeline design
X_5	Sulfate-bearing reducing bacteria (SRB)	X_{18}	Manufacture defects
X_6	Contain erosion	X_{19}	Material defect
X_7	Inner protective layer failure	X_{20}	Construction defect
X_8	Sulphide in soil	X_{21}	Operational mis-operation
X_8	High salinity of soil	X_{22}	Maintenance mis-operation
X_{10}	Low PH of soil	X_{23}	Flood damage
X_{11}	Soil contains corrosive bacteria	X_{24}	Debris flow
X_{12}	High soil redox potential	X_{25}	Landslide
X_{13}	High soil moisture content	X_{26}	Earthquake disaster

3.2. Establish the Factors to Monitor Pipeline Failures

Based on the comprehensive analysis of the failure factor of shale gas gathering and transportation pipelines, the evaluation index factor set is established considering the pipeline condition, transportation medium, natural environmental conditions, and human factors. A factor set is the set of evaluation indexes of a decision-making system: $U = \{u_1, u_2, \ldots, u_n\}$. In the multilevel evaluation model of shale gas pipeline failures, the hierarchical system is divided into three layers: target layer T, parent factor layer, and sub-factor layer. Layer T, the target layer, demonstrates the final objective of the entire hierarchical structure, which is depicted in pipeline failure. The parent factor layer is divided into six types of factors that affect the failure of pipelines, namely third-party damage, internal corrosion, external corrosion, mis-operation, design and construction defects, and natural disasters, which are denoted as M_1, M_2, M_3, M_4, M_5, and M_6. Sub-factor layer is the 26 s-level indices established in the evaluation model, which are shown in Table 6.

The evaluation factor model is established, as shown in Figure 2.

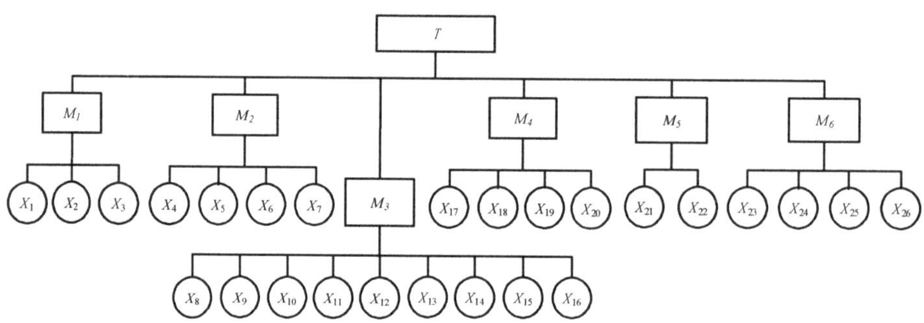

Figure 2. Indicators to monitor pipeline failure.

3.3. Expert Weight Calculation

Based on IFAHP, 10 enterprise experts were invited to score each factor with triangular fuzzy numbers, of which 10 were issued and 10 were recovered (all 10 were valid).

According to those data, the weight and total weight results of each level were obtained. According to the hierarchical data shown in the Figure 1, the judgment matrix of the parent factor layer relative to the target layer is recorded as T, and its eigenvectors are denoted as W_T. The sub-factor layer judgment matrices are recorded as M_1, M_2, M_3, M_4, M_5, and M_6, and the corresponding eigenvectors are recorded as W_1, W_2, W_3, W_4, W_5, and W_6. The total weight result is expressed in W_a. According to Equation (1), the probability of each level is converted into a real number judgment matrix, as shown in Tables 7–13. The calculation results of eigenvectors are as follows:

$W_T = [0.21, 0.18, 0.18, 0.14, 0.14, 0.15]^T$;
$W_1 = [0.36, 0.31, 0.33]^T$;
$W_2 = [0.25, 0.23, 0.26, 0.26]^T$;
$W_3 = [0.11, 0.11, 0.09, 0.10; 0.10; 0.13; 0.12; 0.12; 0.11]^T$;
$W_4 = [0.22, 0.24, 0.23, 0.31]^T$;
$W_5 = [0.50, 0.50]^T$;
$W_6 = [0.25, 0.25, 0.26, 0.24]^T$.

Table 7. Judgment matrix of the parent factor layer to the target layer T.

Parent Factor of T	M_1	M_2	M_3	M_4	M_5	M_6
M_1	1	1	1	1	1	1
M_2	0.66	1	1	1	1	1
M_3	0.60	0.99	1	1	1	1
M_4	0.35	0.35	0.72	1	1	0.98
M_5	0.28	0.62	0.64	0.90	1	0.89
M_6	0.36	0.72	0.73	1	1	1

Table 8. Judgment matrix of the parent factor layer to the target layer M_1.

Sub-Factors of M_1	X_1	X_2	X_3
X_1	1	1	1
X_2	0.75	1	0.92
X_3	0.81	1	1

Table 9. Judgment matrix of the sub-factor layer to the upper factor M_2.

Sub-Factors of M2	X_4	X_5	X_6	X_7
X_4	1	1	0.97	0.94
X_5	0.86	1	0.84	0.80
X_6	1	1	1	0.97
X_7	1	1	1	1

Table 10. Judgment matrix of the sub-factor layer to the upper factor M_3.

Sub-Factors of M_3	X_8	X_9	X_{10}	X_{11}	X_{12}	X_{13}	X_{14}	X_{15}	X_{16}
X_8	1	1	1	1	1	0.46	0.88	0.73	0.99
X_9	0.95	1	1	1	1	0.45	0.84	0.70	0.94
X_{10}	0.78	0.82	1	0.93	0.91	0.26	0.68	0.52	0.77
X_{11}	0.86	0.90	1	1	0.98	0.34	0.75	0.60	0.85
X_{12}	0.88	0.92	1	1	1	0.36	0.77	0.62	0.86
X_{13}	1	1	1	1	1	1	1	1	1
X_{14}	1	1	1	1	1	0.63	1	0.87	1
X_{15}	1	1	1	1	1	0.77	1	1	1
X_{16}	1	1	1	1	1	0.52	0.90	0.76	1

Table 11. Judgment matrix of the sub-factor layer to the upper factor M_4.

Sub-factors of M_4	X_{17}	X_{18}	X_{19}	X_{20}
X_{17}	1	0.91	0.95	0.22
X_{18}	1	1	1	0.37
X_{19}	1	0.96	1	0.27
X_{20}	1	1	1	1

Table 12. Judgment matrix of the sub-factor layer to the upper factor M_5.

Sub-Factors of M_5	X_{21}	X_{22}
X_{21}	1	0.99
X_{22}	1	1

Table 13. Judgment matrix of the sub-factor layer to the upper factor M_6.

Sub-Factors of M6	X_{23}	X_{24}	X_{25}	X_{26}
X_{23}	1	1	0.85	1
X_{24}	0.94	1	0.77	1
X_{25}	1	1	1	1
X_{26}	0.91	0.97	0.75	1

According to the calculation results of the eigenvector, the total weight is calculated as follows.

W_a = [0.07, 0.07, 0.07, 0.05, 0.0423, 0.045, 0.05, 0.02, 0.02, 0.02, 0.02, 0.02, 0.02, 0.02, 0.02, 0.02, 0.03, 0.03, 0.03, 0.04, 0.07, 0.07, 0.04, 0.04, 0.04, 0.04]^T.

3.3.1. Objective Weight Calculation

Considering that regional characteristics and factors are not unified, the Bayesian network model is used to modify, and the formula is used to normalize the probability weight. The event probability refers to the relevant research literature [18–22], and the probability of an intentional destruction factor refers to the literature [23,24]. The Bayesian network diagram is shown in Figure 3.

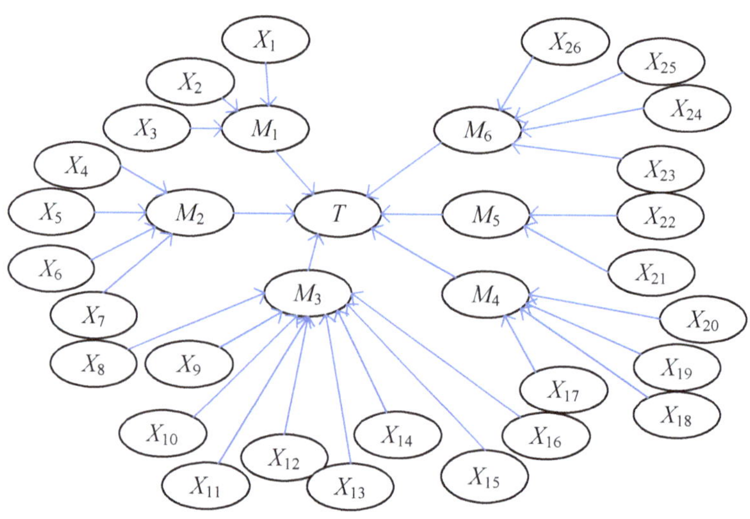

Figure 3. Bayesian network model for pipeline failure factor.

The statistical probability of failure events is also an important basis for determining factor weight. According to OREDA [25], relevant statistics, literature [26–30], and empirical data, the basic probability of events affected by the structure of Bayesian model can be concluded. The posterior probability of Bayesian model can be inferenced according to prior probability. Then, normalize the probability according to the posterior probability of each factor in order to obtain an objective weight relative to minimized subjectivity. Prior probability, posterior probability, and objective weight are shown in Table 14.

Table 14. Pipe failure factors prior probability, posterior probability, and objective weight.

Number	Prior Probability (km·a)	Posterior Probability	Objective Weight	Number	Prior Probability (km·a)	Posterior Probability	Objective Weight
X_1	3.0×10^{-3}	3.2×10^{-2}	3.0×10^{-2}	X_{14}	6.0×10^{-3}	6.4×10^{-2}	6.1×10^{-2}
X_2	1.3×10^{-3}	1.4×10^{-2}	1.3×10^{-2}	X_{15}	5.6×10^{-3}	6.0×10^{-2}	5.7×10^{-2}
X_3	5.0×10^{-3}	5.3×10^{-2}	5.1×10^{-2}	X_{16}	6.5×10^{-3}	7.0×10^{-2}	6.6×10^{-2}
X_4	4.8×10^{-3}	5.1×10^{-2}	4.9×10^{-2}	X_{17}	1.9×10^{-3}	2.0×10^{-2}	1.9×10^{-2}
X_5	6.5×10^{-3}	6.9×10^{-2}	6.6×10^{-2}	X_{18}	1.1×10^{-3}	1.2×10^{-2}	1.1×10^{-2}
X_6	6.6×10^{-3}	7.0×10^{-2}	6.7×10^{-2}	X_{19}	1.1×10^{-3}	1.2×10^{-2}	1.1×10^{-2}
X_7	6.0×10^{-3}	6.4×10^{-2}	6.1×10^{-2}	X_{20}	6.7×10^{-3}	7.2×10^{-2}	6.8×10^{-2}
X_8	4.8×10^{-3}	5.1×10^{-2}	4.9×10^{-2}	X_{21}	2.4×10^{-3}	2.6×10^{-2}	2.5×10^{-2}
X_9	4.5×10^{-3}	4.8×10^{-2}	4.6×10^{-2}	X_{22}	1.8×10^{-3}	1.9×10^{-2}	1.8×10^{-2}
X_{10}	3.0×10^{-3}	3.2×10^{-2}	3.1×10^{-2}	X_{23}	9.0×10^{-4}	0.96×10^{-2}	9.2×10^{-3}
X_{11}	4.8×10^{-3}	5.1×10^{-2}	4.9×10^{-2}	X_{24}	9.0×10^{-4}	0.96×10^{-2}	9.2×10^{-3}
X_{12}	3.5×10^{-3}	3.7×10^{-2}	3.6×10^{-2}	X_{25}	2.2×10^{-3}	2.4×10^{-2}	2.3×10^{-2}
X_{13}	6.0×10^{-3}	6.4×10^{-2}	6.1×10^{-2}	X_{26}	1.2×10^{-3}	1.3×10^{-2}	1.2×10^{-2}

3.3.2. Modification of FAHP by Bayesian Network

In order to make the analysis result more in line with the actual situation, the FAHP method is modified and determined by a Bayesian network. According to the expert scoring weight W_a calculated by IFAHP and the probability weight converted from statistical probability, it is recorded as W_b, and the comprehensive weight is calculated by Formula (3), which is recorded as W_c.

W_c = [0.05, 0.03, 0.06, 0.05, 0.06, 0.06, 0.06, 0.03, 0.03, 0.02, 0.03, 0.03, 0.04, 0.04, 0.04, 0.04, 0.03, 0.02, 0.02, 0.06, 0.04, 0.04, 0.02, 0.02, 0.03, 0.02]^T.

3.4. *Fuzzy Comprehensive Evaluation of Risk in High Consequence Areas*

3.4.1. High Consequence Areas Recognition

According to the identification criteria for high consequence areas of shale gas gathering and transportation pipelines in Table 1, there are three sections of HCAs identified in this pipeline, as shown in Table 5.

3.4.2. Sub-Factor Evaluation

For the identified three high-consequence pipe sections, five experts are invited to adopt the evaluation principle in Table 2 to evaluate the failure factors of each high-consequence pipe section with a risk grade score between 0 and 1, and the sum of membership degrees of each index corresponding to each evaluation grade is 1. Then, five experts were counted to judge the estimated values, and the average value was taken as the final membership estimated values of each index. The evaluation grade and membership degree are shown in Table 15. Only the evaluation results of HCAs1 are listed here, and HCAs2 and HCAs3 methods are consistent.

Table 15. Sub-factors comprehensive weight ranking and membership estimation.

Factor	Weight		Judge Level and Membership				
	Serial Number	Numerical Value	Low (I)	Lower (II)	Medium (III)	Higher (IV)	High (V)
X_3	1	0.063	0.10	0.38	0.28	0.16	0.08
X_6	2	0.062	0.38	0.28	0.22	0.08	0.04
X_7	3	0.060	0.34	0.16	0.42	0.08	0.00
X_{20}	4	0.059	0.36	0.24	0.20	0.12	0.08
X_5	5	0.058	0.44	0.20	0.36	0.00	0.00
X_4	6	0.053	0.14	0.28	0.56	0.02	0.00
X_1	7	0.051	0.48	0.16	0.26	0.08	0.02
X_{21}	8	0.045	0.30	0.38	0.10	0.14	0.08
X_{13}	9	0.041	0.28	0.42	0.14	0.10	0.06
X_{16}	10	0.041	0.20	0.26	0.36	0.12	0.06
X_{14}	11	0.040	0.40	0.30	0.14	0.12	0.04
X_{15}	12	0.039	0.30	0.24	0.32	0.10	0.04
X_{22}	13	0.039	0.14	0.36	0.38	0.08	0.04
X_8	14	0.035	0.20	0.38	0.36	0.06	0.00
X_9	15	0.033	0.18	0.30	0.38	0.10	0.04
X_{25}	16	0.033	0.36	0.28	0.32	0.04	0.00
X_{11}	17	0.033	0.10	0.42	0.36	0.08	0.04
X_2	18	0.031	0.30	0.24	0.36	0.10	0.00
X_{12}	19	0.028	0.32	0.22	0.36	0.06	0.04
X_{17}	20	0.027	0.10	0.30	0.50	0.06	0.04
X_{10}	21	0.025	0.18	0.32	0.42	0.06	0.02
X_{26}	22	0.023	0.30	0.24	0.28	0.12	0.06
X_{18}	23	0.021	0.24	0.40	0.36	0.00	0.00
X_{19}	24	0.021	0.30	0.36	0.24	0.06	0.04
X_{23}	25	0.021	0.18	0.22	0.46	0.08	0.06
X_{24}	26	0.020	0.18	0.56	0.26	0.00	0.00

3.4.3. Multi-Factor Fuzzy Comprehensive Evaluation

Evaluation matrix R is composed of the membership degrees of $X_1 \sim X_{26}$ in Table 15, which is the comprehensive membership degree matrix of sub-factors. The comprehensive weight is W_c. According to Equation (8), the comprehensive evaluation result W_{HCAs1} of pipeline failure sub-factors to the target layer is calculated as follows.

$W_{\text{HCAs1}} = [0.27, 0.29, 0.32, 0.08, 0.04]$.

The comprehensive evaluation results obtained by HCAs2 and HCAs3 are as follows.

$W_{\text{HCAs2}} = [0.26, 0.32, 0.30, 0.08, 0.04]$.

$W_{\text{HCAs3}} = [0.27, 0.36, 0.24, 0.08, 0.04]$.

4. Discussion

According to the principle maximum membership, the risk level of HCAs1 is medium; HCAs2 and HCAs3 in HCAs have lower risk levels. This is roughly consistent with the actual situation in HCAs, during pipeline operation, HCAs1 has been leaked, while HCAs2 and HCAs3 in HCAs have relatively low risks.

The main causes of pipeline failure are M_1, M_2, and M_3. Among them, the failure probabilities of X_1, X_2, X_3, X_4, X_6, and X_7 rank within the top six; therefore, we need to focus on them. They are the weak links in the field; it is, thus, necessary to focus on monitoring and preventing them during the pipelines running through HCAs1. Under the condition of limited resources, priority should be given to X_1, X_2, X_3, and X_7 when developing prevention and control measures.

Therefore, in the actual operation process in addition to routine inspections, the management for HCAs1 should strengthen and improve its safety precautions.

5. Conclusions

An integrated methodology is used to calculate the expert scoring weight of the factors causing the failure possibility of shale gas pipelines by incorporating improved fuzzy analytic hierarchy process and Bayesian updating technique. The advantage of this method is more prominent, which minimizes the subjectivity of experts' scoring weight. The purpose of this study is to classify risk levels in HCAs of shale gas pipeline under the condition of uncertain data. Combined with the probability weight calculated by the statistical probability of the failure possibility factors of shale gas pipelines, the comprehensive weight is calculated to evaluate the failure possibility grade of shale gas pipelines. The study observes that places within 200 m on both sides of the pipeline must be considered as HCAs. This includes drinking water sources, large- and medium-sized rivers, lakes, water reservoirs, nature reserves, and ecological sensitive areas.

This study applies an IFAHP and Bayesian network method for safety risk analysis of pipelines running through HCAs. It is found that third-party damage and protective layer failure account for the largest weight, and design and construction defects account for the lowest weight. This result may be caused by the increase in population density or regional development. Protective measures can be strengthened by using technical means. The final result of the situation is retained by conducting a case study, which verified the viability and effectiveness of the proposed methodology and offers an important reference for hierarchical management of HCAs. Using this method, we can focus on the diagnosis and analysis of the risk factors, which are more likely to cause pipeline leakage, in order to prevent the occurrence of pipeline leakage accidents. The improved FAHP and Bayesian Network method also have some limitations. In the terms of BN model construction processes, they rely on domain experts. Our subsequent research will focus on automatic knowledge acquisition and develop a real-time evaluation and decision support system for shale gas gathering and transportation pipeline risk classification.

Author Contributions: Conceptualization and methodology and software and data curation, N.S. and Z.L.; formal analysis, X.C.; resources investigation, X.W.; writing—original draft preparation, N.S. and X.C.; writing—review and editing, K.C. and S.L.; supervision and project administration, K.C. and W.Q.; funding acquisition, K.C. All authors have read and agreed to the published version of the manuscript.

Funding: This research was funded by The Scientific and Technology Research Program of Chongqing Municipal Education Commission, grant number KJZD-K201901501; Research Foundation of Chongqing University of Science & Technology, grant number YKJCX1820705, YKJCX2020716, YKJCX2020717.

Institutional Review Board Statement: Not applicable.

Informed Consent Statement: Not applicable.

Data Availability Statement: The data presented in this study are available upon request from the corresponding author.

Acknowledgments: The authors are grateful for the Chongqing Administration of Work Safety, the Chongqing Municipal Education Commission and Chongqing University of Science & Technology.

Conflicts of Interest: The authors declare that they have no known competing financial interests or personal relationships that could have appeared to influence the work reported in this paper.

Appendix A Normalization process

According to the scoring results of several different experts combined with different expert information, the comprehensive scoring value of each index is calculated and normalized. The expression is as follows.

$$\overline{R_w} = \sum_{k=1}^{S} \beta_k R_w^k = \left(\sum_{k=1}^{S} \beta_k r_i^k \right)_{1\times n} = \left(\sum_{k=1}^{S} \beta_k l_i^k, \sum_{k=1}^{S} \beta_k m_i^k, \sum_{k=1}^{S} \beta_k u_i^k \right)_{1\times n} = (\overline{r}_i)_{1\times n} \quad \text{(A1)}$$

In the expression, s is the number of experts, $k = 1, 2, \cdots, s$, β is normalized expert weight coefficient, and $\overline{R_w}$ and r_i are the triangular fuzzy number matrix set and the triangular fuzzy number after synthesizing the opinion of experts, respectively.

$$\tilde{S}_i = \frac{\overline{r}_i}{\sum\limits_{i=1}^{n} \overline{r}_i} = \left(\frac{\overline{l}_i}{\sum\limits_{i=1}^{n} \overline{l}_i}, \frac{\overline{m}_i}{\sum\limits_{i=1}^{n} \overline{m}_i}, \frac{\overline{u}_i}{\sum\limits_{i=1}^{n} \overline{u}_i} \right) \quad \text{(A2)}$$

In the expression, $\widetilde{S}_i$ is the normalized value of index C_i, the normalized comprehensive judgement matrix is S_w, and the expression is as follows.

$$S_w = \begin{bmatrix} \tilde{S}_1 & \tilde{S}_2 & \cdots & \tilde{S}_n \end{bmatrix}_{1\times n} \quad \text{(A3)}$$

References

1. Chen, X.; Wu, Z.; Chen, W.; Kang, R.; He, X.; Miao, Y. Selection of key indicators for reputation loss in oil and gas pipeline failure event. *Eng. Fail. Anal.* **2019**, *99*, 69–84. [CrossRef]
2. Li, M.; Zheng, H.; Xue, X.; Xue, L.; Ai, M.; Ma, W. Reliability evaluation and management of PetroChina's large-scale system of natural gas pipeline networks. *J. Nat. Gas Geosci.* **2019**, *4*, 287–295. [CrossRef]
3. Stephens, M.; Leewis, K.G.; Moore, D.K. A Model for Sizing High Consequence Areas Associated With Natural Gas Pipelines. In Proceedings of the 2002 4th International Pipeline Conference, Calgary, AB, Canada, 29 September–3 October 2002. [CrossRef]
4. Onisawa, T. An approach to human reliability in man-machine systems using error possibility. *Fuzzy Sets Syst.* **1988**, *27*, 87–103. [CrossRef]
5. Lam, C.; Zhou, W. Statistical analyses of incidents on onshore gas transmission pipelines based on PHMSA database. *Int. J. Press. Vessel. Pip.* **2016**, *145*, 29–40. [CrossRef]
6. Zhang, P.; Peng, X.; Li, X. Oil and Gas Pipeline Limited Charge Optimum Maintenance Decision-Making Analysis Based on Fuzzy-Gray-Element Theory and Fuzzy Analytical Hierarchy Process. In Proceedings of the International Conference on Fuzzy Information and Engineering South West Petroleum University, Chengdu, China, 13–16 May 2007. [CrossRef]
7. Wang, X.; Shuai, J.; Song, H.; Wang, Y. Identification and hierarchical management of high consequence area for oil pipeline. *China Saf. Sci. J.* **2015**, *25*, 149–154. [CrossRef]
8. Dong, S.; Wang, D.; Fei, F.; An, Y.; Dong, Q.; Zhou, Y. Upgrading of pipeline regions and control of public security risks. *Chin. J. Oil Gas Storage Transp.* **2014**, *11*, 1164–1170. [CrossRef]
9. Shan, K.; Shuai, J. Study on management of risk in natural gas pipelines after upgrading location class. *China Saf. Sci. J.* **2016**, *26*, 145–150. [CrossRef]
10. Yao, A.L.; Zhou, L.G.; Wang, L.; Wang, X.T.; Li, Y.L. Management of and risk evaluation on long-distance gas pipelines related to regional level upgrading. *Chin. J. Nat. Gas Ind.* **2017**, *37*, 124–130. [CrossRef]
11. Kim, J.; Lee, J.; Kim, B.C.; Kim, J. Raw material criticality assessment with weighted indicators: An application of fuzzy analytic hierarchy process. *Resour. Policy* **2019**, *60*, 225–233. [CrossRef]
12. Guo, C.; Khan, F.; Imtiaz, S. Copula-based Bayesian network model for process system risk assessment. *Process Saf. Environ. Prot.* **2019**, *123*, 317–326. [CrossRef]
13. Khakzad, N.; Khan, F.; Amyotte, P. Dynamic safety analysis of process systems by mapping bow-tie into Bayesian network. *Process Saf. Environ. Prot.* **2013**, *91*, 46–53. [CrossRef]
14. Lee, Y.; Haberman, S.; Dorans, N. Use of Adjustment by Minimum Discriminant Information in Linking Constructed-Response Test Scores in the Absence of Common Items. *J. Educ. Meas.* **2019**, *56*, 452–472. [CrossRef]
15. Haberman, S.; Yao, L. Repeater Analysis for Combining Information from Different Assessments. *J. Educ. Meas.* **2015**, *52*, 223–251. [CrossRef]
16. *GB 32167*; Oil and Gas Pipeline Integrity Management Specification. China National Standardization Administration Committee: Beijing, China, 2015.
17. Halim, S.; Yu, M.; Escobar, H.; Quddus, N. Towards a causal model from pipeline incident data analysis. *Process Saf. Environ. Prot.* **2020**, *143*, 348–360. [CrossRef]
18. Yan, C.; Quddus, N.; Mashuga, C. Bayesian Network and Game Theory Risk Assessment Model for Third-Party Damage to Oil and Gas Pipelines. *Process Saf. Environ. Prot.* **2019**, *134*, 178–188. [CrossRef]
19. Guo, Y.; Meng, X.; Wang, D.; Meng, T.; Liu, S.H.; He, R.Y. Comprehensive risk evaluation of long-distance oil and gas transportation pipelines using a fuzzy petri net model. *J. Nat. Gas Sci. Eng.* **2016**, *33*, 18–29. [CrossRef]
20. Wang, X.; Duan, Q. Improved AHP–TOPSIS model for the comprehensive risk evaluation of oil and gas pipelines. *Pet. Sci.* **2019**, *16*, 245–258. [CrossRef]
21. Li, X.; Chen, G.; Zhu, H. Quantitative risk analysis on leakage failure of submarine oil and gas pipelines using Bayesian network. *Process Saf. Environ. Prot.* **2016**, *103*, 163–173. [CrossRef]

22. Badida, P.; Balasulbramaniam, Y.; Jayapraakash, J. Risk evaluation of oil and natural gas pipelines due to natural hazards using fuzzy fault tree analysis. *J. Nat. Gas Sci. Eng.* **2019**, *66*, 284–292. [CrossRef]
23. Guo, X.; Zhang, L.; Liang, W.; Haugen, S. Risk identification of third-party damage on oil and gas pipelines through the bayesian network. *J. Loss Prev. Process Ind.* **2018**, *54*, 163–178. [CrossRef]
24. Zarei, E.; Yazdi, M.; Abbassi, R.; Khan, F. A hybrid model for human factor analysis in process accidents: Fbn-hfacs. *J. Loss Prev. Process Ind.* **2019**, *57*, 142–155. [CrossRef]
25. Sintef, N.T.N.U. *Offshore and Onshore Reliability Data Handbook*; OREDA Participants: Oslo, Norway, 2015.
26. Wang, T.; Wang, X.; Zairong, L.; Xue, L.; Gao, Z.; Wang, Y. Comparison on failures of long- distance oil & gas pipelines at home and abroad. *Oil Gas Storage Transp.* **2017**, *36*, 1258–1264. [CrossRef]
27. Shan, K.; Shuai, J.; Yang, G.; Meng, W.; Zhang, H. Evaluation method for basic failure probability of oil and gas pipeline based on accident statistics in the United States and enlightenments. *Chin. J. Integr. Reliab.* **2020**, *39*, 530–535. [CrossRef]
28. Huang, P.; Xu, J.; Lai, Y.; Zhang, Y. Human factors analysis of leakage explosion of oil and gas pipeline based on HFACS and AHP. *Saf. Environ. Eng.* **2016**, *23*, 114–118. [CrossRef]
29. Liu, J.; Zhu, Y.; Qin, H.; Liu, W.; Xiao, W. Risk analysis of subsea x-tree system leakage based on bayesian networks. *Lubr. Eng.* **2018**, *43*, 109–114. [CrossRef]
30. Hao, Y.; Xing, Z.; Wang, K.; Shao, H.; Wei, J. Quantitative risk analysis of natural gas pipeline failure. *Oil Gas Storage Transp.* **2011**, *30*, 263–265.

processes

MDPI

Article

Bubble Identification in the Emerging Economy Fuel Price Series: Evidence from Generalized Sup Augmented Dickey–Fuller Test

Mumtaz Ahmed [1], Muhammad Irfan [1], Abdelrhman Meero [2], Maryam Tariq [1], Ubaldo Comite [3], Abdul Aziz Abdul Rahman [2], Muhammad Safdar Sial [4,*] and Stefan B. Gunnlaugsson [5]

1 Department of Economics, COMSATS University Islamabad (C.U.I.), Islamabad 44000, Pakistan; mumtaz.ahmed@comsats.edu.pk (M.A.); mirfan@comsats.edu.pk (M.I.); tariqmariam96@gmail.com (M.T.)
2 College of Business Administration, Kingdom University, Riffa 40434, Bahrain; ar.meero@ku.edu.bh (A.M.); a.abdulrahman@ku.edu.bh (A.A.A.R.)
3 Department of Business Sciences, University Giustino Fortunato, 82100 Benevento, Italy; u.comite@unifortunato.eu
4 Department of Management Sciences, COMSATS University Islamabad (C.U.I.), Islamabad 44000, Pakistan
5 Department of Business Administration, University of Akureyri, 600 Akureyri, Iceland; stefanb@unak.is
* Correspondence: safdarsial@comsats.edu.pk

Abstract: In the recent past, the world in general and Pakistan in particular faced a drastic fuel price change, affecting the economic productivity of the country. This has drawn the attention of empirical researchers to analyze the abrupt change in fuel prices. This study takes a lead and investigates for the first time, in the literature related to Pakistan, the presence of multiple fuel price bubbles, with the purpose of knowing if the price driver is due to demand or it is exuberant consumer behavior that prevails and contributes to a sudden boom in fuel price series. The empirical analysis is performed through a recently proposed state-of-the-art generalized sup ADF (GSADF) approach on six commonly used fuel price series, namely, LDO (light diesel oil), HSD (high-speed diesel), petrol, natural gas, kerosene, and MS (motor spirit). The bubble analysis for each of the six fuel price series is based on monthly data from July 2005 to August 2020. The findings provide evidence of the existence of multiple bubbles in all series considered. Specifically, four bubbles are detected in each of the kerosene and natural gas price series, whereas three bubbles are noted in each of the HSD, LDO, petrol and MS price series. The maximum duration of occurrence of bubbles is of 12 months for kerosene. The date-stamping of the bubbles shows that the financial crisis of 2008 contributed to the emergence of bubbles that pushed oil prices upward and caused a depreciation in the national currency.

Keywords: price bubbles; bubble length; petroleum products; GSADF; Pakistan

Citation: Ahmed, M.; Irfan, M.; Meero, A.; Tariq, M.; Comite, U.; Abdul Rahman, A.A.; Sial, M.S.; Gunnlaugsson, S.B. Bubble Identification in the Emerging Economy Fuel Price Series: Evidence from Generalized Sup Augmented Dickey–Fuller Test. *Processes* **2022**, *10*, 65. https://doi.org/10.3390/pr10010065

Academic Editor: Jean-Claude Assaf

Received: 1 December 2021
Accepted: 24 December 2021
Published: 29 December 2021

1. Introduction

A bubble can be defined as a rapid increase in the price of a good, which increases due to the exuberant behavior of the consumer or the market. Bubbles appear in the market when the investor invests in market assets, securities, goods and so forth, in the interest of profitability. This rapid price increase is much more than the productive capacity or the actual value of an asset that no one on the market wants to buy. This leads to a sharp decline in the price of an asset where everyone begins to sell their assets, creating economic turbulence [1].

With the increase in output, oil is considered as an indicator that has determined the economic development of a country. Oil is considered of prime importance for each country, so its price affects the production of a country. An increase in the price of oil leads to an increase in the cost of production, import bills, and price of petroleum goods, and thus a reduction in the production capacity of a country due to high input cost, which results in low demand, low investment as well as poor economic growth and lower purchasing power [2].

Caspi and Katzke [3] detected a bubble in US crude oil from the WTI monthly price series using the generalized sup ADF (ADF) test procedure. The recursive identification algorithms used in this analysis recognize several price exclusivities with macro variables. Zhang and Yao [4] inspected the presence of bubbles in the crude oil, diesel and petrol markets using the State–Space model and the log-periodic power law (LPPL) model. Su and Li [5] detected the existence of multiple bubbles in West Texas Intermediate (WTI) crude oil price by using the generalized sup Augmented Dickey–Fuller (GSADF) method and found six bubbles from 1986 to 2016.

Perifanis [6] examined West Texas Intermediate (WTI) price bubbles, as well as the length and causes of the bubble. The results of this paper indicate that there are speculative episodes in the Brent oil price series, but not in the WTI price series. Li and Chevallier [7] detected the presence of a bubble in the US, European, and Asian natural gas markets by using the generalized sup ADF (GSADF) test, and the results indicate the presence of bubbles. All of this literature review shows that there have been studies in the past that have found multiple bubbles in the oil price series.

Thus, the following subsections show some historical bubbles that the world has witnessed, which have been clarified as Tulip Mania, the South Sea Bubble, the Dotcom Bubble, and the US Housing Bubble. Going back to history, the most significant speculative bubble was the Tulip Mania that began in 1636 and early 1637 in Holland. Tulips of different colors were introduced and people began to buy them in abundance. The demand overwhelmed the supply of tulips to such an extent that its price reached a point where it was sold at a cost where people mortgage their homes and businesses.

However, by the end of 1637, after reaching a peak, the price of tulips had fallen and left tulip owners bankrupt. The South Sea Bubble, which had destroyed many investors in 1720, was because of a joint stock company that was established in 1711 in London, supplying slaves for Spanish plantations in Central and South America. In anticipation of the success of East India Company, investors took the stock of South Sea Company. There was an implausible boom in the company's shares in 1720, but stock prices reached much higher than the company's profits, which drove the market to collapse in September. The company's stocks plummeted destroying many investors.

Due to the increasing popularity of the Internet in the 1990s, many investors invested in Internet-based startups in the hope of earning profits. The confidence of investors that their investments are going to generate profits created an environment, the NASDAQ composite index, which contained shares of dot.com businesses, and which reached more than 5000 in 2000. However, the index fell quickly because investors neglected the traditional basic investment matrices and caused a recession in the United States. Many economists believe that due to the bursting dot.com bubble, many investors concluded that investing in real estate was safer and better. They began investing in housing markets that doubled housing prices in the United States between 1996 and 2006. In 2006, the value of homes in the US increased and then began to fall, resulting in the average US home losing one-third of its value in 2009. American homes soared and crashed, and the impact on mortgage-backed securities resulted in the first global economic crisis since the Great Depression.

The Asian crisis that took place worldwide in 1997 affected Singapore, Malaysia, Indonesia, Taiwan, the Philippines and South Korea. Later, there was the global financial crisis in 2007–2008, when many financial institutions went bankrupt. The failure of these institutions resulted in a global financial market freeze and required government intervention on a global scale. The European debt crisis and unemployment in Greece and Spain have had a major negative impact on the economy and the labor market. The collapse of price bubbles on financial markets, such as foreign exchange markets, equity markets and housing markets, has caused all these crises.

These events highlight the importance of bubble detection and have revealed its implications. It has been shown that the bursting of a bubble can lead to the collapse of large financial institutions, bring countries to the brink of bankruptcy and lead to complete financial and economic crises. Consequently, the detection of a bubble is very important because a sudden eruption can cause a disruption of economic activity, which leads to other disrupting factors such as unemployment, low wages, financial instability, and the disruption of economic activity.

Most developed and developing countries produce electricity through alternate sources such as renewables including wind, solar, water, etc.; however, Pakistan depends on oil for the production of electricity, which is quite an expensive input. An increase in the oil prices impacts the GDP of Pakistan, contributes to inflation, a rise in the budget deficit, and makes import expensive by putting downward pressure on the exchange rate. Much work is available on bubble detection for the oil prices considering data of different countries; however, no empirical work is available on bubble detection in commonly used fuels for Pakistan. The present research takes a lead and fills this void, addressing this very significant issue in Pakistan using the newly developed Phillips and Shi [8] approach—the generalized supremum ADF (GSADF). The approach has several advantages over other strategies generally used to detect bubble detection as it can detect periodically collapsing multiple bubbles, whereas previous approaches were designed to detect only a single bubble. The GSADF approach also provides the information on start and end dates of bubbles by using its date-stamping strategy. Thus, the present study will be a good addition to the existing literature, particularly for Pakistan.

The remainder of the paper is organized as follows. Section 2 provides a thorough and critical summary of the literature. Section 3 looks at the theoretical and econometric methodology. Section 4 contains the results and discussions, while Section 5 presents some relevant policy proposals and the conclusion of the study.

2. Literature Review

The literature on bubble detection exists not only in oil price series, but also in various financial markets, such as gold, agriculture, stocks, housing, etc. See, for example, Campbell and Lo [9], Diba and Grossman [10], and Chang and Aye [11] for stock markets; Brooks and Prokopczuk [12] for housing markets; Fantazzini [13] for oil prices; Flood and Garber [14] for inflation; Caspi and Katzke [3] for commodity price bubbles.

With reference to the oil market, studies have been undertaken in different economies covering WTI, BRENT, US, Asia, Europe, and many others, by employing different methods of testing, such the usual right-tailed Augmented Dickey–Fuller (ADF) test, sup-ADF and the generalized SADF tests.

Hall and Psaradakis [15] detect the collapse of the consumer price bubble and Argentina's exchange rate in terms of monthly USD data. Time series data from 1983 to 1989 were tested by using the Dickey–Fuller test and Markov switching approach. The results revealed that it is better to switch to the ADF approach to identify bubbles in the exchange rate market; therefore, the null hypothesis, which endorses explosive behaviors in the data, is rejected.

Yumashev and Ślusarczyk [16] speculated the presence of bubbles in the US stock market by applying a momentum threshold autoregressive (MTAR) model, a non-linear time series technique to detect bubbles. Time series from 1871 to 1999, together with monthly data from 1871 to 2001, were used. The results of the analysis of two subsamples of data (1871 to 1995) and a subsample of data (1871 to 2001) confirm the existence of bubbles. General findings conclude that the market collapsed as a result of the presence of a bubble in the 1990s.

In their study, An and Mikhaylov [17] developed a model which made it possible to quantitatively determine whether a bubble exists in the gold market or not. They calculated the difference between the market and the fundamental price. Data were taken from August 2011 to September 2011 to determine if there exists a martingale or a local martingale under the risk-neutral possibilities through random process. If the volatility is too large as measured through a modified Florens–Zmirou estimator then there exists a bubble. However, the results concluded that gold price is a martingale and not a local martingale, so there exists no bubble in the gold market during the testing period.

Fantazzini [13] argued that in 2014/2015, there was a negative bubble in oil prices, reducing them above the extent warranted by economic fundamentals. Two sets for bubble prediction methods have been corroborated: the first set consists of financial bubble measures, while the second set consists of the log-periodic power law (LPPL) model for negative financial bubbles. The data of daily nominal and real WTI (West Texas Intermediate) oil prices series were taken from January 2013 to April 2015. This time frame was selected because the researcher concentrated on the price collapse that occurred at the end of 2014. The results conclude that in the last months of 2014 and early 2015, the price of oil experienced a financial bubble.

In their paper based on historical crude oil, Zhang and Yao [4] analyzed diesel and gasoline market data taken from November 2001 to December 2015. The study used the State–Space model and the log-periodic power law (LPPL) model to detect dynamic oil price bubbles and predict the moment of their collapse. The results conclude that, first, there are only oil market bubbles between November 2001 and July 2008, and bubbles are the driving force behind the rise in crude oil and diesel prices. Secondly, the State–Space model detects the bubbles in crude oil and diesel prices that differ in time. Finally, the crash time of bubbles is well estimated by the LPPL model.

In their paper, Dinçer and Lisin [18] examined the existence of multiple bubbles in West Texas Intermediate (WTI) crude oil price by using the generalized sup Augmented Dickey–Fuller (GSADF) method proposed by Phillips and Shi [8]. Crude oil is considered to be of primary importance in both the energy sector and the capital market. Data were collected from January 1986 to April 2016 and multiple bubbles on the WTI oil market were detected in 1990, 2005, 2006, 2008 and 2015. The results indicate that there were six bubbles between 1986 and 2016 and that oil prices differ from their inherent value. The bubble dates are consistent with particular events in the global and financial markets.

Caspi and Katzke [3] detected the existence of multiple bubbles by using the generalized sup ADF (GSADF) test procedure developed [8]. The researchers used monthly data from 1876 to 2014 and classified explosive cycles of oil prices in comparison to the general market level and stocks of oil in the US after 1876 and 1920, respectively. The recursive identification algorithms used in this analysis recognize several periods of price exclusivity and provide guidance for examining the macroeconomic effect of historic periods with large increases in oil prices.

Liaqat and Nazir [19] used the GSADF to detect the existence of numerous bubbles at the Pakistan Stock Exchange across different industrial sectors. Monthly data from 2007 to 2016 were used for the analysis and indicated that the industries where there are no stock bubbles are investment, chemicals and textile spinning. They concluded in their analysis that several bubbles were present in the KSE-100 index.

Perifanis [6] researched for a bubble in West Texas Intermediate (WTI) from the periods (1 January 1947 and 1 September 2018). The main purpose of the study, besides the identification of the existence and duration of the bubbles, was to understand the cause behind each explosive bubble. The study used two methodologies including PWY (2011) and PSY (2015) for bubble detection. The PWY (2011) date-stamping process detects only two bubble periods in real WTI prices. In comparison, PSY (2015) suggests seven bubbles, while the era of the 1978 oil crisis is implicated as an explosion period.

In their paper, Liz and Ruiz-Herrera [20] examined the presence of multiple bubbles for WTI and BRENT oil price series by taking weekly data between 1990 and 2019 using GSADF right tail tests initiated. The result of this paper indicates the presence of speculative episodes in Brent oil price series but not in the WTI price series because the former is a more united market and less vulnerable to speculative movements.

Fatima and Ahmed [21] tested for the existence of bubbles in the agricultural commodities of Pakistan. They used monthly data of agricultural commodities from January 2000 to May 2018 (wheat, rice, cotton, sugar, maize, barley, and soybean) by applying the GSADF method. The analysis revealed the presence of 22 bubbles in agricultural commodities. Butt and Ahmed [22] took the initiative to detect the presence of multiple bubbles in five distinct sets of inflation rates such as the CPI, SPI, WPI, CPIG and CPIF by applying GSADF. The observational results were built on data taken from the monthly time series from January 2006 to January 2019. The results led to the conclusion that a number of bubbles were detected in the non-food WPI and CPI. For the remaining series, only a single bubble has been detected.

Li and Chevallier [7] detected the origin and termination of housing bubbles in 140 US cities during the period from 1989 to 2017. Many factors were examined in this study: first, the existence of housing bubbles in US cities; second, the identification of the period of housing bubbles and the determination of the origin of housing bubbles; and third, the potential basis of regional variation in the existence of bubbles. The methodology used for bubble detection is GSADF (2013) and the results of the study indicate bubbles in 40 cities out of 140.

Li and Chevallier [7] detected bubbles in all three potential US, European and Asian markets. Monthly data from January 1996 to June 2017 were taken for analysis and it was found that the price bubble of the natural gas industry has a great impact on the economy at micro and macro levels. The detection of gas price bubbles is necessary in order to develop effective strategies to deal with the explosion of bubbles. The results indicate the presence of bubbles between January 1996 and June 2017 in the European market, January 1996 and June 2017 in the Asian market, and between January 1996 and June 2017 in the US market.

Li and Chevallier [7] examined volatility in asset prices due to the existence of bubbles. He used the real stock prices and dividend data from 1871 to 1997, consisting of 122 observations. The methodology used was augmented Dickey–Fuller tests. The results of the findings conclude that much of the deviation from the actual prices was due to the presence of bubbles and the H0 hypothesis, which states that no bubble is rejected as the conclusion shows that the deviation is frequently attributed to the presence of bubbles in the model, which deviates the actual asset prices.

Hall and Psaradakis [15] detected the collapsing bubble in consumer prices and the exchange rate of Argentina in terms of USD monthly data. Time series data for the analysis was taken from 1983 to 1989. Data were tested using the Dickey–Fuller test and by employing the Markov switching approach. The results show that it is better to switch to the ADF approach to identify bubbles in the exchange rate market; therefore, the null hypothesis, which endorses explosive behaviors in the data, is rejected.

Brooks and Prokopczuk [12] determine the occurrences of bubbles in the UK for traded equity property stocks. The time series data have been taken from 1986 to 1998 by employing a co-integration analysis. The methodology includes the dividend discount and Gordon growth method as well as variance bound tests, which are used to detect the presence of bubbles. However, the results of the findings conclude that bubbles existed in the 1980s, but that no bubbles existed in the 1990s. Inclusively, these bubbles are short-lived.

Bohl [23] speculated the presence of bubbles in the US stock market by applying a momentum threshold autoregressive (MTAR) model, a non-linear time series technique to detect bubbles that cause sudden rise followed by an abrupt crash. Time series data from 1871 to 1999, as well as monthly data from 1871 to 2001, are analyzed. The results of the analysis in two subsamples of data (1871 to 1995) suggested that there was no major bubble, and a subsample of data (1871 to 2001) implied the existence of bubbles since the 1990s, so the general findings conclude that the market crashed due to the presence of a bubble.

In their paper, Case and Shiller [24] analyzed whether there exist multiple bubbles in the housing market or not. Their study was conducted through a survey supervised in 2003, of the people of four big cities (Los Angeles, San Francisco, Boston, and Milwaukee) that bought houses. A random sample method was used for the survey. The results of the survey show that several factors affect housing buying behavior and also tell us whether a housing bubble exists or not. These factors include investment motives and the desire for additional price increase, the need for urgency to purchase a home, and adherence to short-sighted hypotheses about the housing markets. However, the results showed that there existed no bubble in 2003 as there was in 1988; the 2003 survey was a replica of the 1988 survey, with the difference that Milwaukee was added as a new case in 2003. These results also indicate that these prices cannot be sustained at such high levels for long and that a bubble is inevitable.

In their paper, Noussair and Tucker [25] considered future markets and the formation of a bubble in the asset market, which differed from Smith et al.'s (1988) study where they observed bubbles and crashes. The results of the study indicated that when future markets are present, bubbles do not occur in the spot market because the future market decreases the speculation and decision errors that seem to give rise to price bubbles in asset markets. The data was taken from October 2002 to April 2003, involving 12 traders in each of the sessions.

Wang and Fan [26] detected the rational bubble in the Korean stock market. The data was taken from 1996 to November 2007; they used three integration tests, namely, the JJ, KSS, and BN approaches. The findings of the traditional JJ test confirmed the existence of rational bubbles, while both the nonlinear KSS test and the nonparametric BN test proved that rational bubbles may not have been present in the capital market of Korea.

In their paper, Jarrow and Kchia [27] developed a model that can quantitatively determine whether a bubble exists in the gold market or not. They calculated the difference between the market and the fundamental price. The data were taken from 25 August 2011 to 1 September 2011. They analyzed it to determine if there existed a martingale or a local martingale under the risk-neutral possibilities. If the volatility was too large as measured through a modified Florens–Zmirou estimator, then there existed a bubble; however, the results indicated that gold price is a martingale and not a local martingale, so there existed no bubble in the gold market during the testing period.

Baur and Glover [28] established a study to determine bubbles in gold prices. They used the method developed by Phillips and Shi [8]. They used different start dates and then they were able to detect periods of significant and insignificant explosive price behaviors. From 2002 to 2012, the results of the estimates indicate significant explosive price behaviors; however, in 2008, this behavior was disturbed by the financial crisis. However, the results show strong evidence of the bubble in gold prices.

Yiu and Yu [29] detected bubbles in the residential property market of Hong Kong. Data was taken from March 1993 to March 2011 and the recently built PSY approach which defines the asset price bubble in the residential real estate of the Hong Kong property market. The findings show nine bubbles between 1994 and 2011, and the result shows the speculative bubble in the Hong Kong property market in 1997, another positive bubble from January to March 2004, which was short-lived, and lastly, our approach distinguishes two bubbles, one from April to August 2009 and one from February to March 2011.

In their paper, Bouoiyour and Selmi [30] investigated whether Bitcoin is an income or a speculative bubble. Their study's main objective was to determine the main driving force behind the Bitcoin price, business transaction, and consumer attractiveness. Data used for Bitcoin price was taken on a daily basis through the blockchain website (ETR); the user's attractiveness to Bitcoin was collected through daily Bitcoin views on Wikipedia. However, the results uncovered a few contrasts regarding the frequencies in question, featuring the multi-faced nature of surveying what Bitcoin resembles and more trouble to identify this developing cryptocurrency. This paper concludes that Bitcoin is a speculative bubble without disregarding its convenience for monetary reasons.

Etienne and Irwin [31] tested the presence of bubbles in US future agricultural markets. The data was taken from 1970 to 2011 and an ADF test was applied for testing. The results concluded that multiple price explosiveness exists and most of the bubbles are momentary and last less than ten days; moreover, the results also suggested that approximately one-third of the bubbles are negative. All these factors include incompetent commodity markets, increase trading volume, and technological advancement.

Adämmer and Bohl [32] investigated corn, soya bean, and wheat prices in the US. Data were taken from January 1983 to April 2011. The results indicate that bubbles exist in wheat prices because it is a high energy-consuming commodity as compared to others and has a large share in US exports, while for corn and soya bean, the results were inconclusive.

de Holanda Barbosa and da Silva Filho [33] tested bubble detection for hyperinflation in Brazil. They used the time series tax revenue data from 1947 to 2003, using the ADF approach. The results indicated that strong hyperinflation is rejected whereas weak hyperinflation is not rejected and the hypothesis that hyperinflation is caused due to bubble presence is rejected.

Coskun and Jadevicius [34] detected bubbles in Turkey's housing market by using the GSADF approach. Data was taken on a monthly basis from January (2010) to June (2014). The results of the observations oppose the existence of bubbles over the chosen duration of time and also suggest that the theory that demand conditions have led prices to emerge is true.

Greenaway-McGrevy and Phillips [35] examined the detection of housing bubbles in the metropolitan centers in New Zealand. Therefore, they used data from 1993 to 2014, by applying the bubble tests initiated by Phillips et al. The results indicate that the real estate sector in Auckland City contributed to the rise of bubbles in the other centers by up to two quarters in 2003, as well as other TAs in the Auckland area by a single quarter, which was due to high cost of real estate

Gomez-Gonzalez and Gamboa-Arbeláez [36] examined the bubble detection in housing prices for 18 OECD countries. Data was taken quarterly from 1970 to 2013. The methodology used was GSADF and the co-explosive technique suggested by Engsted and Nielsen [37] to test for bubble detection. The results of the study show the explosiveness in housing prices which supports the hypothesis of the bubble presence.

In their paper, Bevilacqua and Fassò [38] analyzed whether the Internet is facing another bubble, keeping in mind the 1999–2000 dotcom bubble. He put forward four criteria to determine whether there exists a bubble or not. The first criteria consist of the determination of venture capital funding as it is the main culprit of a speculative bubble. The second consists of initial public offerings. The third was the NASDAQ stock exchange as it played a major role in the dotcom bubble phenomena. The last was to analyze different external signals. However, the results concluded that the venture capital funding is very high and the majority of the portion was taken by the technological industry, and soon after, a downward trend showed that a bubble might be present. As far as IPO earning is concerned, the data showed the companies went public before earing any profits as they went public at the time of dotcom bubble, which shows that bubble might have existed on its way to the tech company. As the external signals are concerned, they concluded that according to their results, the venture capital index has declined, which may not confirm a bubble; however, if there exists a bubble, it might burst very soon.

Arshanapalli and Nelson [39] detected bubbles through four econometric tools. The variance bound test was used but, due to several errors in the test, a new test was developed, named the West Test. The West Test could detect bubbles but it could not identify the start and end dates. Another test called co-integration was used but it faced the same problem as the West Test, so they then used a unit root test that was capable of detecting bubbles present; however, it still could not detect multiple bubbles. To detect multiple bubbles, the ADF test was developed by Phillips and Shi [8] which detects multiple bubbles as well as detecting the start and end dates of a bubble. However, the author of the paper concluded that the accessibility of such real-time monitoring instruments would encourage financial specialists and portfolio directors to rebalance their portfolios during the existence of such bubbles in the stock market.

Caramugan and Bayacag [40] analyzed exports, mainly rice, rubber, and palm oil, by using monthly time series data from 1980 to 2015. They used GSADF and date stamps for the initiation and end of bubbles. The results of the finding concluded that there exist multiple bubbles; numerous factors have backed price swells such as low stock, rising oil prices, weak currency, and instability in commodity markets.

Alexakis and Bagnarosa [41] tested bubble detection in the prices of hog corn and soybean future meals. They used monthly time series data from January 2001 to April 2016. Data series were analyzed using the BSDAF approach and co-integrated vector autoregressive model. However, the results concluded that bubbles in feed do not co-integrate with the price of the hog.

Greenaway-McGrevy and Phillips [35] established a study to identify bubbles in house prices-to-rent ratios in Australian capital cities. To investigate whether the house price-to-rent ratio in the national capital, Canberra, and any of the state capital cities (Adelaide, Brisbane, Hobart, Melbourne, Perth, and Sydney) demonstrated explosive activity at any point in time, they used the PSY research algorithm. The data used from December 1995 to August 2015 were sourced from SIRCA's (2015) Core Logic RP online database. The findings show proof of asset bubbles in many capital cities, many of which are broadly consistent with the construction boom in the first half of the 2000s.

Liaqat and Nazir [19] used the GSADF model to detect the existence of numerous bubbles in the Pakistan stock exchange through various industrial sectors. They used monthly data from 2007 to 2016. The analysis reveals that the only industries in which no stock bubbles are found are investment, chemical, and textile spinning, and concluded in their analysis the presence of several bubbles in the KSE-100 Index, along with numerous industrial sectors.

Zhang and Wang [42] detected multiple bubbles in the Chinese food market. Monthly data were taken from January 1990 to December 2017 using the GSADF technique. The results indicate that four explosive bubbles episodes prevailed. These bubbles occurred due to various factors such as supply and demand differences, depreciation in USD, financial crisis, and speculation.

Fatima and Ahmed [21] tested for the existence of bubbles in agricultural commodities of Pakistan. This study led to the detection of bubbles in different agricultural commodities of Pakistan. The statistics used were monthly data for the price of agricultural commodities from January 200 to May 2018 (wheat, rice, cotton, sugar, maize, barley, and soybean) by using the newly developed, state-of-the-art GSADF method developed by Phillips et al. (2015). The analysis indicates that in all the price sequences, 22 bubbles were found in total.

In their paper, Wei and Li [43] detected housing bubbles in 140 US cities and also endeavored to understand the bubbles' origin and termination, during the period from 1989 to 2017. Many factors are examined in this study: first, the existence of housing bubbles in US cities; second, the identification of the period of housing bubbles and the determination of the origin of housing bubbles; and third, the potential basis of regional variation in the existence of bubbles. The methodology used for bubble detection is (GSADF 2013) and the results of the finding suggest 40 cities out of 140 cities where there is a bubble.

Elsayed and Danial [44] detected the presence of multiple bubbles in five separate series of inflation rates including CPI, SPI, WPI, CPIG, and CPIF. They used the recently developed methodology of GSADF. The observational results were built on data taken from the monthly time series of January 2006 to January 2019. However, the findings concluded that several bubbles were found in non-food WPI and CPI. For the remaining three series, only a single bubble was detected.

Table 1 provides details of the studies that are performed specifically on oil price series across the world.

Table 1. Summary of existing studies on oil price series.

S. No.	Author	Country	Time Period	Data Series	Methodology	Results
1	Caspi et al. (2015)	United States	1876–2014	Oil prices	GSADF	Bubbles in the US
2	Fantazzini (2016)	United States WTI	January 2013 to April 2015	Oil prices	LPPL	Negative financial bubble
3	Zhang and Yao (2016)	United States WTI	November 2001–December 2015	Crude oil, diesel, and gasoline	state-space model and log-periodic power law (LPPL)	Bubble detection in all three markets
4	Sua et al. (2017)	U.S. WTI	January 1986 to April 2016	Crude oil price	GSADF	six bubbles from 1986–2016
5	Perifanis (2019)	U.S. WTI	January 1947 and September 2018	Oil prices	PWY (2011) and PSY (2015),	Two bubbles detected by PWY and seven bubbles by PSY
6	Christian et al. (2019)	U.S. WTI and BRENT	1990 and 2019	Oil price	SADF and GSADF right tail	Bubble in BRENT oil market but not in WTI
7	Li et al. (2020)	U.S, Europe, Asia	January 1996 to June 2017	Natural gas	GSADF	Bubble exists in all three markets

From the extensive literature review, it is concluded that some studies have been conducted on bubble detection in other relevant areas, such as those by Liaqat, Nazir [19] Butt and Ahmed [22] but no studies have been carried out on detecting oil prices in Pakistan. Most of the papers reviewed used the methodology that were capable of detecting a single bubble and were not able to identifying the exact date or hidden bubbles. These drawbacks led us to use the latest methodology available to detect the presences of multiple bubbles.

In this study, however, an attempt is made to fill this gap by detecting the presence of multiple bubbles in Pakistan's fuel price series. The analytical results from this paper will help investors and policy makers understand fuel price volatility of HSD, LDO, MS RON, natural gas, petrol, and kerosene in the presence of a bubble. The results of the study will be used by decision-makers to make future economic decisions aimed at preventing financial risk and maintaining financial stability.

3. Methodology

3.1. Theoretical Background

The development of the commodity price bubble is widely certified by speculation and the chaos of trading entities. Various theories relate to this phenomenon such as commodity price theory, animal spirit, the greater fool theory, and extrapolation theory [45]. These theories suggest that when economic agents are irrational, they distort the value of assets, resulting in market volatility. All of these theories are rooted in the Keynesian concept of irrational behavior.

The most commonly used theoretical framework developed by Pindyck and Rotemberg [46] is a rational theory of commodity prices. Within this framework, present value is applied to rational commodity prices where p_t is taken as the commodity cost that is considered by present and expected future payments denoted by y_t + 1. It is highlighted that y_t + 1 is the convenience yield in this theory for storable item adjustments that accumulate inventory as profits which, in turn, can be earned through deals and stock out avoidance, whereas the convenience yield is the premium associated with possessing an underlying commodity rather than related derivative security.

The general arbitrage condition is:

$$P_t = E_t\left[\sum_{i=1}^{t-1} \frac{1}{(1+D)^i}(y_t+1)\right] + E_t\frac{1}{(1+D)^{t-1}}P_t \tag{1}$$

where D is the discount rate, The commodity price is P_t, and $Y_t + 1$ is the profit earned from sales of output or commodity.

The price of the commodity is identified by time 't' through demand and supply, whereas the second term in Equation (1) is bubble component that is diverged by market fundamentals. If there is no existence of a bubble, then:

$$\lim_{N\to\infty} E_t\left[\frac{(P_t+N)}{(1+D)^N}\right] = 0 \tag{2}$$

This shows that when a bubble does not exist the price of commodities solely reflects the fundamental dynamics of the market, then Equation (1) becomes:

$$P_t^f = \sum_{j=1}^{\infty} \frac{1}{(1+D)^i} E_t[y_{t+j}] \tag{3}$$

Equation (3) is known as the transversality condition. It means that without bubbles, the fundamentals of the market depend solely on the price of the goods. Without the implication of a transversality, the condition price of a commodity at t is simplified as:

$$P_t = F_t + B_t \tag{4}$$

Here, F_t is the fundamental component, and B_t is a bubble component.

When $B_t = 0$ then P_t is defined by market fundaments F_t and P_t, with an integrated process of order one. When F_t is integrated of order then P_t follows the same integrated process.

3.2. Econometric Methodology

The approach employed in this research builds on the work undertaken by Phillips and Shi [8] that was developed to overcome the weaknesses of the SADF approach developed [47]. The problem with SADF is that it can identify the existence of one bubble but cannot detect multiple bubbles. To overcome this issue, GSADF has been developed to detect the existence of multiple bubbles as well as the length of time. The regression model that explains the SADF and GSADF process is as follows:

$$Q_t = \rho + \varpi Q_{t-1} + \sum_{k=1}^{n} \theta_k \Delta Q_{t-k} + \epsilon_t \tag{5}$$

where Q_t denotes the index price series of (petrol, HSD, LDO, kerosene, natural gas, MS), ρ is intercepted, ϖ is the coefficient of the first lag of Q_t, coefficient of ΔQ_{t-k} is θ_k, and error at time 't' with mean zero and constant variance is ϵ_t.

The objective of this study is to find explosive behavior in oil price series which is completed by framing the null hypothesis as

$$H_0\colon \varpi = 1 \tag{6}$$

Contrary to the right-tailed alternative:

$$H_1\colon \varpi > 1 \tag{7}$$

Some notations have been added to understand the discussion. Firstly, we have standardized the sample to change over into [0, 1] range. Let $\varpi_{s_{1,s_2}}$ and $ADF_{s_{1,s_2}}$ denote the estimated coefficient of Q_{t-1} in [1] and the equivalent ADF statistic over the standardized sample $[s_1, s_2]$. Further, let λ_s be the window size represented by $\lambda_s = s_2 - s_1$.

Before we explain the GSADF and SADF tests, it is important to understand the right-tailed version of the unit root test. We consider s_1, s_2 the first and last observation of the chosen sample, and the window size for example is $\lambda_s = 1$ which suggests that the critical value of RTADF will be different from the standard ADF. When the estimated value of RTADF is compared with the corresponding critical values which is 1%, 5%, and 10%. If the measured value is found to be greater than the critical value, then we reject the null hypothesis and conclude the existence of a bubble

The sup-ADF elaborates the ADF statistics with constant initial point and fluctuating window size, i.e., λ_s. The window size is chosen as $0.01 + 1.8/\sqrt{T}$ which is proposed by Philips et al. (2015). In the estimation of the window size, S_1 is considered to be a starting point that is $S_1 = 0$ and S_2 being the endpoint that is $S_2 = 1$. According to negligible window size S_0 which is mentioned earlier as initial window size $\lambda_s = S_2$. Regression estimation is achieved by augmenting the window size $s_2 \in [s_0, 1]$. An ADF statistics (ADF_{s_2}) is determined for each estimation by keeping one observation at a time. However, the final estimation is carried on the whole sample (i.e., $s_2 = 1$).

$$\text{SADF}\ (s_0) = \underbrace{sup}_{s_2 \in [s_0, 1]} \{ADF_{S2}\} \tag{8}$$

The generalized version of SADF that is the GSADF is widely accepted because of its flexible variable window size. In this approach, the starting window size can adjust within the given range of $[0, s_1 - s_0]$

$$\text{GSADF}\ (s_0) = \underset{\substack{s_2 \in [s_0 - 1] \\ s_1 \in [0,\ s_2 - s_0]}}{Sup} \{\text{ADF}_{s_1}^{s_2}\} \tag{9}$$

3.3. Date Stamping of Bubble

SADF and GSASF are used to date stamp bubbles, as both tests can systematically determine the start and end period of a bubble. We can estimate the start and finish of a bubble by rejecting the null hypothesis of one of these tests. To determine the starting and ending point of each bubble at the time T_{r2}, we have to compare each calculated value of ADF_{s2} with its corresponding standard right-tailed ADF critical values. The starting point $(s_{st\hat{a}rt})$ of the bubble is when ADF_{s2} is greater than the critical value $(ADF_{s2} > CV)$; this is the point of bubble emergence and the ending point $(s_{e\hat{n}d})$ is when ADF_{s2} is lesser than the critical value $(ADF_{s2} < CV)$. Therefore, bubble estimates are defined as:

$$s_{st\hat{a}rt} = \underbrace{inf}_{s_2 \in [s_0, 1]} \{s_2 : ADF_{s_2} > crit_{s_2}^{\beta T}\} \tag{10}$$

$$s_{e\hat{n}d} = \underbrace{inf}_{s_2 \in [s_{st\hat{a}rt}\ ,1]} \left\{s_2 : ADF_{s_2} < crit_{s_2}^{\beta T}\right\} \tag{11}$$

where the critical value of the ADF is $crit_{w_2}^{\beta T}$ is $100\ (1 - \beta T)\%$ which is based on observation T_{r2}. We can determine the bubble period of the GSADF procedure in a similar way which is as follows:

$$s_{st\hat{a}rt} = \underbrace{inf}_{s_2 \in [s_0, 1]} \{s_2 : BSADF_{s_2}(s_0) > crit_{s_2}^{\beta T}\} \tag{12}$$

$$s_{e\hat{n}d} = \underbrace{inf}_{s_2 \in [s_{st\hat{a}rt}\ ,1]} \left\{s_2 : BSADF_{s_2}(s_0) < crit_{s_2}^{\beta T}\right\} \tag{13}$$

where the critical value of the sup ADF statistics is $crit_{s_2}^{\beta T}$ is the 100 $(1 - \beta_T)$% built on observations $[T_{r2}]$. The backward sup ADF statistic is BSADF (s_0) for $s_2 \in [s_0, 1]$, that describes GSADF statistic denoted by

$$\text{GSADF } (s_0) = \underset{s_2 \in [s_0, 1]}{sup} [BSADF_{s_2}(s_0)] \tag{14}$$

3.4. Data and Its Sources

This study considers the monthly price series from January 2010 to October 2020 for MS (motor spirit), LDO (light diesel oil), HSD (high-speed diesel), and kerosene oil for bubble detection, while for petrol and natural gas, data from July 2005 to February 2017 is used. Crude oil and other related oils are not considered due to the non-availability of data or large number of missing values. Data have been collected through different sources such as PSO, OGDCL and ORGA. Since there were fewer data available for natural gas and gasoline compared to other series, a bubble comparison can be made to the kerosene, HSD, LDO and MS. Table 2 shows the list of variables along with their description and measuring unit.

Table 2. Variable details and measuring units.

S. No.	Variable Name	Measuring Unit
1	Natural gas	PKR per cubic feet
2	Petrol	PKR per liter
3	MS (motor spirit)	PKR ml
4	LDO (light diesel oil)	PKR centistoke
5	HSD (high-speed diesel)	PKR centistoke
6	Kerosene	PKR per liter

4. Results and Discussion

This section provides details of the empirical results covering basic summary statistics of the data series under consideration followed by the results of GSADF test and finally presenting date-stamping of bubbles in each of the fuel price series with information on the bubble start and end dates. The detailed discussion of results is also provided. Table 3 presents descriptive statistics on the six oil price series (petrol, HSD, LDO, MSRON, natural gas, and kerosene). Table 4 presents the RTDF, SADF and GSADF statistics for the six series and rejects the null hypothesis of no bubble in the series. Table 5 shows the stamping periods of the six oil price series of the selected period, their start and end dates as well as their duration.

Table 3. Descriptive statistics.

	MS	PETROL	KERO	LDO	HSD	Natural Gas
MEAN	79.294	90.91	66.338	62.385	81.380	85.75
SD	19.566	26.488	24.673	22.87	27.885	15.886
MEDIAN	74.8	80.52	62.25	60.00	82.0	83.85
IQR	34.26	40.13	46.02	40.18	43.92	29.07
SKEWNESS	0.3496	0.71498	0.14169	0.36459	0.2244	0.15295
KURTOSIS	1.865	2.088	1.5917	1.6786	1.9525	1.7632
MIN	48.94	54.33	29.53	27.84	31.74	60.27
MAX	117.83	143.9	108.13	101.24	132.467	115.33

Note: authors own calculation.

Table 4. Results of GSADF statistics for oil prices series.

S. No.	OIL PRICE	RTADF	SADF STATISTICS	GSADF STATISTICS
1	Natural gas	3.00 *** (0.00)	0.29 *** (0.37)	3.00 *** (0.00)
2	HSD	4.50 *** (0.00)	3.83 *** (0.00)	5.45 *** (0.00)
3	Kerosene	3.99 ** (0.00)	3.71 *** (0.00)	5.04 *** (0.00)
4	MS RON	3.41 *** (0.00)	3.51 *** (0.00)	4.36 *** s(0.00)
5	LDO	4.59 *** (0.00)	4.56 *** (0.00)	6.19 *** (0.00)
6	Petrol	4.80 *** (0.00)	3.09 *** (0.00)	6.41 *** (0.00)

Note: The *p*-value of each statistic is presented in parentheses and **, ***, respectively denotes significance at 5% and 1% significance level.

Table 5. Bubble date stamping of oil process.

Commodity	Number of Bubbles	Start Date	End Date	Duration (Months)
Natural gas	4	February 2012	June 2012	5
		OCTOBER 2014	April 2015	7
		December 2017	May 2018	6
		May 2018	July 2018	3
HSD	3	February 2008	November 2008	10
		November 2014	March 2015	5
		January 2018	July 2018	7
MS	3	March 2008	August 2008	6
		November 2014	March 2015	5
		May 2018	July 2018	3
Petrol	3	March 2008	October 2008	8
		February 2012	April 2012	3
		November 2014	May 2015	7
LDO	3	March 2008	November 2008	9
		November 2014	July 2015	9
		April 2018	November 2018	8
Kerosene	4	March 2008	June 2008	4
		June 2008	October 2008	5
		November 2014	May 2015	7
		January 2018	December 2018	12

Note: authors own calculation.

Table 5 illustrates that the greatest number of bubbles occurred in the natural gas and kerosene price series. The date stamp for each price series was noted and indicates that four bubbles were detected in the price of natural gas from February 2012 to June 2012 for the duration of five months. The second bubble was from October 2014 to April 2015, the third bubble period was from December 2017 to May 2018 and the fourth bubble started from May 2018 to July 2018. Three bubbles were found in the HSD oil price series with the longest duration of eleven months between February 2008 and November 2008. Likewise, for the petrol price series, three bubbles were observed from March 2008 to October 2008 and the longest bubble was eight months. In the LDO, three bubbles were detected, the first bubble was between March 2008 and November 2008 for a duration of nine months. The second bubble occurred from November 2015 to July 2015 and the last bubble seen in the LDO is from April 2018 to November 2018. Finally, in kerosene, the longest bubble was observed during twelve months between January 2018 and December 2018. A total of five bubbles were observed in the kerosene oil price series.

The analysis indicates that, with the exception of natural gas, the other five oil prices experienced a bubble in 2008 and it was a year of global recession. Oil prices surged and the trend accelerated after the recession, as Pakistan is an import economy and dependent on oil imports which influences the overall price of the country. In 2008, BRENT oil prices rose significantly in January from USD 92/bbl to USD 139.83/bbl in June 2008 (Bloomberg). The depreciation of the Pak rupee in dollars has led to very expensive imports and the upward trend in oil prices has doubled the impact on domestic petroleum products. Both reasons paved the way for a rise in oil prices.

The fluctuation of oil prices has a profound impact on the overall economic situation. The supply and demand conditions in the oil market that determine oil prices. Price volatility may be due to three reasons: (a) oil supply shocks normally occur in oil-exporting countries such as the Middle East due to lack of investment and geopolitical tensions, (b) due to exceptionally strong economic growth in developing countries or a relatively rapid economic rebound in some countries in the aftermath of the financial crisis, (c) the specific demand for oil caused by the temporal importance of oil among alternative energy sources and the shifting perception of oil fundamentals. There is a consensus in the literature that the oil demand shocks that were triggered by world economic activity were the major drivers of the bubble in Pakistan's fuel price series.

Figure 1a–f shows the general behaviors of a single bubble and multiple bubbles during the duration of the sample and its rise and fall. To find a particular bubble, the measured value of backward SADF is compared to the critical value. When the computed value exceeds the critical value, the bubble exists and otherwise collapses. The graph shows the start and endpoint clearly. This method is encouraging because it can detect multiple bubbles and stamping the date of bubbles can be easily defined.

The results of our research are not directly comparable to existing studies because of differences in the selected country and data, and the methodology employed in this research is not the same as in many other studies. However, as a comparison, research has been conducted on the detection of the oil price bubble. Caspi and Katzke [3] detected the existence of multiple bubbles in US oil stocks by using the generalized sup ADF (GSADF) test procedure developed by Phillips, Shi [8]. The methodology used allows the identification of bubbles that break down periodically. Fantazzini [13] detected a negative financial bubble in WTI oil prices. Zhang and Yao [4] detected a bubble in three financial markets using the LPPL method. Su and Li [5] detected six bubbles in WTI oil prices using a similar approach to that which we used in our paper (GSADF). Herrera and Tourinho [20] also detected a bubble in BRENT oil prices using GSADF. Li amd Chevallier [7] detected bubbles in three financial markets of Europe, Asia, and the US using the GSADF method. However, Herrera and Tourinho [20] rejected bubble detection in WTI oil prices.

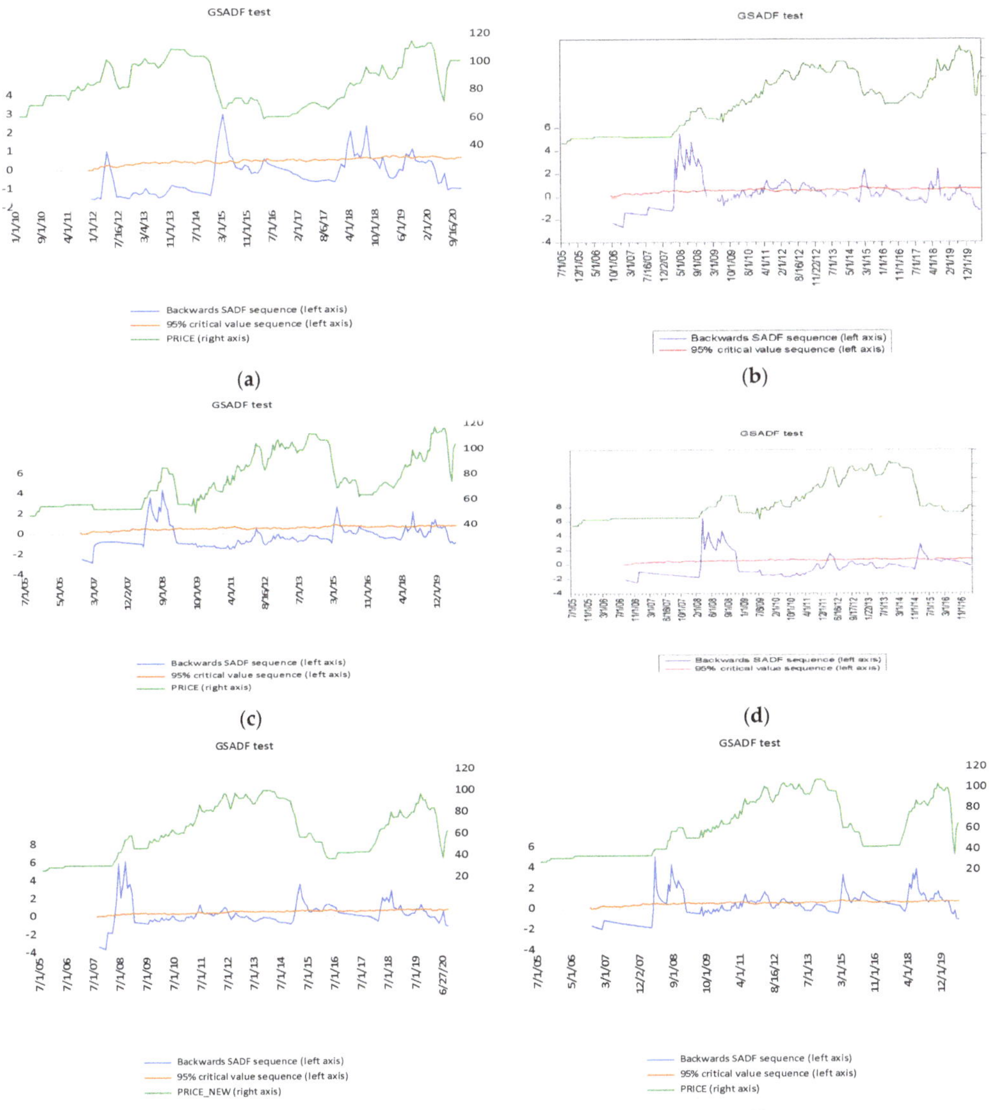

Figure 1. Bubble date stamping of fuel price series. (**a**) Bubble date stamping of natural gas, (**b**) Bubble date stamping of HSD, (**c**) Bubble date stamping of MS RON, (**d**) Bubble date stamping of petrol, (**e**) Bubble date stamping of LDO, (**f**) Bubble date stamping of kerosene. The green line in each sub-graph (**a**–**f**) indicates fuel price series.

5. Conclusions and Policy Recommendation

A rise in oil prices is usually assumed to fuel inflation and limit economic growth; as far as inflation is concerned, oil prices have a strong impact on the price of goods based on oil. The trend in oil prices is increasing globally and domestically as it is in Pakistan. In Pakistan, the immediate problem of oil imports stems from foreign exchange reserves. The continued rise in foreign oil prices has adversely affected Pakistan's balance of payments and fiscal position. In 2008, Pakistan had the worst external deficit and a GDP inflation rate of around 17 percent. The reason for the price increase is due to higher sales in anticipation of potential demand and currency depreciation. For the reasons mentioned, it is important to determine what changes are happening in the oil price series in Pakistan.

Pakistan is heavily dependent on the import of oil and petroleum products to facilitate its consumption needs, and this dependence hurts the economy when an unexpected event occurs. The 1973 oil financial crisis, the 2007/2008 financial crisis and the Great Recession negatively affected Pakistan's economy. Not only does the financial crisis affect the economy, but natural disasters such as the 2005 earthquake in Pakistan have shaken the whole country and triggered inflation in just about every sector of the country. In 2011, Pakistan suffered from major floods that destroyed economic stability.

These events led to higher import prices and a shortage of domestic-produced goods. Most economies generate electricity from alternative sources such solar, wind, hydro, etc. but Pakistan relies on oil for its electricity production, which is a fairly expensive input. All these issues must be addressed because they directly affect oil prices that affect macroeconomic factors and create turmoil in the economy. Pakistan is rich in natural resources and has great geographic importance; the Government of Pakistan should take advantage of this by developing policies that ensure economic improvement.

More importantly, Pakistan should reduce its dependence on oil and petroleum-related products, and use the resources available to it. It should depend on other sources of production because world oil prices directly affect domestic oil prices and can lead to the formation of a bubble. Oil consumption is critical for any economy; it can fuel growth, but considerable attention is necessary for price regulation against oil shocks. A policy plan should be considered by the regulator which is able to minimize oil price declines due to panic sales because petroleum markets have an important impact on the non-energy commodity market. Furthermore, the regulator should consider the possibility of both a negative and a positive bubble.

The only limitation of the current research is unavailability of relevant time series data on other important fuels commonly used in Pakistan such as LPG (liquefied petroleum gas) and CNG (compressed natural gas). Future research can be carried out by incorporating these two fuels as wells and also using longer time series data. The same analysis can be extended for other countries.

Author Contributions: All of the authors contributed to conceptualization, formal analysis, investigation, methodology, and writing and editing of the original draft. All authors have read and agreed to the published version of the manuscript.

Funding: This research received no external funding.

Institutional Review Board Statement: Not applicable.

Informed Consent Statement: Not applicable.

Data Availability Statement: The data will be made available on request from the corresponding author.

Conflicts of Interest: The authors declare no conflict of interest.

References

1. Berger, A.N.; Humphrey, D.B. Measurement and efficiency issues in commercial banking. In *Output Measurement in the Service Sectors*; University of Chicago Press: Chicago, IL, USA, 1992; pp. 245–300.
2. Loungani, P. Oil price shocks and the dispersion hypothesis. *Rev. Econ. Stat.* **1986**, *68*, 536–539. [CrossRef]
3. Caspi, I.; Katzke, N.; Gupta, R. Date stamping historical periods of oil price explosivity: 1876–2014. *Energy Econ.* **2018**, *70*, 582–587. [CrossRef]
4. Zhang, Y.-J.; Yao, T. Interpreting the movement of oil prices: Driven by fundamentals or bubbles? *Econ. Model.* **2016**, *55*, 226–240. [CrossRef]
5. Su, C.-W.; Li, Z.-Z.; Chang, H.-L.; Lobonţ, O.-R. When will occur the crude oil bubbles? *Energy Policy* **2017**, *102*, 1–6. [CrossRef]
6. Perifanis, T. Detecting west Texas intermediate (WTI) prices' bubble periods. *Energies* **2019**, *12*, 2649. [CrossRef]
7. Li, Y.; Chevallier, J.; Wei, Y.; Li, J. Identifying price bubbles in the US, European and Asian natural gas market: Evidence from a GSADF test approach. *Energy Econ.* **2020**, *87*, 104740. [CrossRef]

8. Phillips, P.C.; Shi, S.; Yu, J. Testing for multiple bubbles: Historical episodes of exuberance and collapse in the S&P 500. *Int. Econ. Rev.* **2015**, *56*, 1043–1078.
9. Campbell, J.Y.; Lo, A.; MacKinlay, C. *The Econometrics of Financial Markets*; Princeton University Press: Princeton, NJ, USA, 1997.
10. Diba, B.T.; Grossman, H.I. Explosive rational bubbles in stock prices? *Am. Econ. Rev.* **1988**, *78*, 520–530.
11. Chang, T.; Aye, G.C.; Gupta, R. *Testing for Multiple Bubbles in the BRICS Stock Markets*; Working Paper Series; University of Pretoria Department of Economics: Pretoria, South Africa, 2014; Volume 7.
12. Brooks, C.; Prokopczuk, M.; Wu, Y. Booms and busts in commodity markets: Bubbles or fundamentals? *J. Futures Mark.* **2015**, *35*, 916–938. [CrossRef]
13. Fantazzini, D. The oil price crash in 2014/15: Was there a (negative) financial bubble? *Energy Policy* **2016**, *96*, 383–396. [CrossRef]
14. Flood, R.P.; Garber, P.M. Market fundamentals versus price-level bubbles: The first tests. *J. Political Econ.* **1980**, *88*, 745–770. [CrossRef]
15. Hall, S.G.; Psaradakis, Z.; Sola, M. Detecting periodically collapsing bubbles: A Markov-switching unit root test. *J. Appl. Econom.* **1999**, *14*, 143–154. [CrossRef]
16. Yumashev, A.; Ślusarczyk, B.; Kondrashev, S.; Mikhaylov, A. Global indicators of sustainable development: Evaluation of the influence of the human development index on consumption and quality of energy. *Energies* **2020**, *13*, 2768. [CrossRef]
17. An, J.; Mikhaylov, A.; Richter, U.H. Trade war effects: Evidence from sectors of energy and resources in Africa. *Heliyon* **2020**, *6*, e05693. [CrossRef]
18. Dinçer, H.; Lisin, A.; Ubay, G.G.; Çağlayan, Ç. Identifying the Best Financing Sources for Renewable Energy Companies in Latin American Countries. In *Strategic Approaches to Energy Management*; Springer International Publishing: Cham, Switzerland, 2021; pp. 1–12.
19. Liaqat, A.; Nazir, M.S.; Ahmad, I. Identification of multiple stock bubbles in an emerging market: Application of GSADF approach. *Econ. Change Restruct.* **2019**, *52*, 301–326. [CrossRef]
20. Liz, E. and A. Ruiz-Herrera, The hydra effect, bubbles, and chaos in a simple discrete population model with constant effort harvesting. *J. Math. Biol.* **2012**, *65*, 997–1016. [CrossRef] [PubMed]
21. Fatima, H.; Ahmed, M. Testing for Exuberance Behavior in Agricultural Commodities of Pakistan. 2019. Available online: https://mpra.ub.uni-muenchen.de/95304/ (accessed on 29 November 2021).
22. Butt, M.D.; Ahmed, M. Testing for Multiple Bubbles in Inflation for Pakistan. 2019. Available online: https://mpra.ub.uni-muenchen.de/96847/ (accessed on 29 November 2021).
23. Bohl, M.T. Periodically collapsing bubbles in the US stock market? *Int. Rev. Econ. Financ.* **2003**, *12*, 385–397. [CrossRef]
24. Case, K.E.; Shiller, R.J. Is there a bubble in the housing market? *Brook. Pap. Econ. Act.* **2003**, *2003*, 299–362. [CrossRef]
25. Noussair, C.; Tucker, S. Do Informal Sanctions Increase Cooperation in the Long Run? In *Econometric Society 2021 Australasian Meetings*; Econometric Society, University of Melbourne: Melbourne, Australia, 2021.
26. Wang, Y.; Fan, W.; Liu, Y.; Zeng, Z.; Hao, X.; Chang, M.; Zhang, C.; Xu, Y.; Xiang, H.; Li, Y. Modeling of the Fischer–Tropsch synthesis in slurry bubble column reactors. *Chem. Eng. Processing Process Intensif.* **2008**, *47*, 222–228. [CrossRef]
27. Jarrow, R.; Kchia, Y.; Protter, P. How to detect an asset bubble. *SIAM J. Financ. Math.* **2011**, *2*, 839–865. [CrossRef]
28. Baur, D.G.; Glover, K.J. Speculative trading in the gold market. *Int. Rev. Financ. Anal.* **2015**, *39*, 63–71. [CrossRef]
29. Yiu, M.S.; Yu, J.; Jin, L. Detecting bubbles in Hong Kong residential property market. *J. Asian Econ.* **2013**, *28*, 115–124. [CrossRef]
30. Bouoiyour, J.; Selmi, R.; Tiwari, A. *Is Bitcoin Business Income or Speculative Bubble? Unconditional vs. Conditional Frequency Domain Analysis*; MPRA Paper No. 59595; University Library of Munich: Munich, Germany, 2014.
31. Etienne, X.L.; Irwin, S.H.; Garcia, P. Bubbles in food commodity markets: Four decades of evidence. *J. Int. Money Financ.* **2014**, *42*, 129–155. [CrossRef]
32. Adämmer, P.; Bohl, M.T. Speculative bubbles in agricultural prices. *Q. Rev. Econ. Financ.* **2015**, *55*, 67–76. [CrossRef]
33. de Holanda Barbosa, F.; da Silva Filho, T.N.T. Bubble, weak and strong hyperinflation: Theory and empirical evidence. *EconomiA* **2015**, *16*, 145–156. [CrossRef]
34. Coskun, Y.; Jadevicius, A. Is there a housing bubble in Turkey. *Real Estate Manag. Valuat.* **2017**, *25*, 48–73. [CrossRef]
35. Greenaway-McGrevy, R.; Phillips, P.C. House Prices and Affordability. *N. Z. Econ. Pap.* **2021**, *55*, 1–6. [CrossRef]
36. Gomez-Gonzalez, J.E.; Gamboa-Arbeláez, J.; Hirs-Garzón, J.; Pinchao-Rosero, A. When bubble meets bubble: Contagion in OECD countries. *J. Real Estate Financ. Econ.* **2018**, *56*, 546–566. [CrossRef]
37. Engsted, T.; Nielsen, B. Testing for rational bubbles in a coexplosive vector autoregression. *Econom. J.* **2012**, *15*, 226–254. [CrossRef]
38. Bevilacqua, M.; Fassò, A.; Gaetan, C.; Porcu, E.; Velandia, D. Covariance tapering for multivariate Gaussian random fields estimation. *Stat. Methods Appl.* **2016**, *25*, 21–37. [CrossRef]
39. Arshanapalli, B.; Nelson, W.B. Testing for stock price bubbles: A review of econometric tools. *Int. J. Bus. Financ. Res.* **2016**, *10*, 29–42. [CrossRef]
40. Caramugan, K.M.; Bayacag, P. Price Bubble in Selected ASEAN Agricultural Exports: An Application of the Generalized Supremum Augmented Dickey Fuller. 2016. Available online: https://mpra.ub.uni-muenchen.de/74807/ (accessed on 29 November 2021).
41. Alexakis, C.; Bagnarosa, G.; Dowling, M. Do cointegrated commodities bubble together? the case of hog, corn, and soybean. *Financ. Res. Lett.* **2017**, *23*, 96–102. [CrossRef]

42. Zhang, D.; Wang, T.; Shi, X.; Liu, J. Is hub-based pricing a better choice than oil indexation for natural gas? Evidence from a multiple bubble test. *Energy Econ.* **2018**, *76*, 495–503. [CrossRef]
43. Wei, Y.; Li, Y.; Li, J.; Wang, Y.; Qiang, Z. Regional and longitudinal disparity of housing bubbles in US markets: Evidence from GSADF tests. *J. Urban Plan. Dev.* **2020**, *146*, 04019027. [CrossRef]
44. Elsayed, E.A.; Danial, E.N.; Wadaan, M.A.; El-Enshasy, H.A. Production of β-galactosidase in shake-flask and stirred tank bioreactor cultivations by a newly isolated Bacillus licheniformis strain. *Biocatal. Agric. Biotechnol.* **2019**, *20*, 101231. [CrossRef]
45. Jiménez, Á.J. "Understanding Economic Bubbles", Programa Universitat-Empresa, Barcelona. 2011. Available online: www.eco.uab.es/ue/trabajos%20premi/tfc%2061%20Jim%C3%A9nez%201.pdf (accessed on 10 June 2014).
46. Pindyck, R.S.; Rotemberg, J.J. The comovement of stock prices. *Q. J. Econ.* **1993**, *108*, 1073–1104. [CrossRef]
47. Phillips, P.C.; Wu, Y.; Yu, J. Explosive behavior in the 1990s Nasdaq: When did exuberance escalate asset values? *Int. Econ. Rev.* **2011**, *52*, 201–226. [CrossRef]

MDPI
St. Alban-Anlage 66
4052 Basel
Switzerland
Tel. +41 61 683 77 34
Fax +41 61 302 89 18
www.mdpi.com

Processes Editorial Office
E-mail: processes@mdpi.com
www.mdpi.com/journal/processes

www.ingramcontent.com/pod-product-compliance
Lightning Source LLC
Chambersburg PA
CBHW061512210926
51850CB00008B/1893